D1489324

WITHDRAWN

Energy Alternatives

by Andrea C. Nakaya

Current Issues

ReferencePoint Press™

San Diego, CA

© 2008 ReferencePoint Press, Inc.

For more information, contact
ReferencePoint Press, Inc.
PO Box 27779
San Diego, CA 92198
www.ReferencePointPress.com

Picture credits:
Maury Aaseng, 35–38, 53–56, 71–74, 88–92
Photos.com, 10, 15

Series design:
Tamia Dowlatabadi

LIBRARY OF CONGRESS CATALOGING-IN-PUBLICATION DATA

Nakaya, Andrea C., 1976–
    Energy alternatives / by Andrea C. Nakaya.
        p. cm. — (Compact research)
    Includes bibliographical references and index.
    ISBN-13: 978-1-60152-017-3 (hardback)
    ISBN-10: 1-60152-017-4 (hardback)
    1. Renewable energy sources—Juvenile literature. I. Title.
    TJ808.2.N34 2008
        333.79'4—dc22

                                                                        2007018025

Contents

Foreword

66 **Where is the knowledge we have lost in information?** 99

—"The Rock," T.S. Eliot.

As modern civilization continues to evolve, its ability to create, store, distribute, and access information expands exponentially. The explosion of information from all media continues to increase at a phenomenal rate. By 2020 some experts predict the worldwide information base will double every 73 days. While access to diverse sources of information and perspectives is paramount to any democratic society, information alone cannot help people gain knowledge and understanding. Information must be organized and presented clearly and succinctly in order to be understood. The challenge in the digital age becomes not the creation of information, but how best to sort, organize, enhance, and present information.

ReferencePoint Press developed the *Compact Research* series with this challenge of the information age in mind. More than any other subject area today, researching current events can yield vast, diverse, and unqualified information that can be intimidating and overwhelming for even the most advanced and motivated researcher. The *Compact Research* series offers a compact, relevant, intelligent, and conveniently organized collection of information covering a variety of current and controversial topics ranging from illegal immigration to marijuana.

The series focuses on three types of information: objective single-author narratives, opinion-based primary source quotations, and facts and statistics. The clearly written objective narratives provide context and reliable background information. Primary source quotes are carefully selected and cited, exposing the reader to differing points of view. And facts and statistics sections aid the reader in evaluating perspectives. Presenting these key types of information creates a richer, more balanced learning experience.

For better understanding and convenience, the series enhances information by organizing it into narrower topics and adding design features that make it easy for a reader to identify desired content. For example, in *Compact Research: Illegal Immigration*, a chapter covering the economic impact of illegal immigration has an objective narrative explaining the various ways the economy is impacted, a balanced section of numerous primary source quotes on the topic, followed by facts and full-color illustrations to encourage evaluation of contrasting perspectives.

The ancient Roman philosopher Lucius Annaeus Seneca wrote, "It is quality rather than quantity that matters." More than just a collection of content, the *Compact Research* series is simply committed to creating, finding, organizing, and presenting the most relevant and appropriate amount of information on a current topic in a user-friendly style that invites, intrigues, and fosters understanding.

Energy Alternatives at a Glance

World Energy Consumption

The majority of the world's energy is generated by fossil fuels, and the amount of energy generated increases every day. The World Coal Institute projects that global demand will increase 60 percent by 2030.

America's Energy Supply

The United States consumes a disproportionate amount of energy. According to the Energy Information Administration it has less than 5 percent of the world's population but consumes more than 20 percent of its energy.

Reliance on Fossil Fuels

Some people believe society cannot continue to rely on fossil fuels because reserves will soon run out. Others contend that the world still has vast coal and natural gas supplies and that new oil reserves will be discovered. The International Energy Agency estimates that coal reserves will last for 164 years.

America's Use of Alternatives

While the United States is increasing its use of alternative energies such as ethanol and wind power, other nations are doing so more rapidly. Iceland generates all of its electricity from renewable sources, and ethanol makes up about 40 percent of Brazil's automobile fuel.

Nuclear Power

Some people advocate nuclear power as a nonpolluting, highly concentrated energy alternative to fossil fuels. It currently provides about 20 percent of America's electricity; however, safety issues and radioactive waste disposal are big concerns.

Energy Alternatives

Solar power, wind power, hydropower, geothermal energy, and ocean energy are all potential energy alternatives. However, they still supply only a small amount of the total energy consumed; in 2004 only about 6 percent of the United States' energy came from these sources.

Alternative Transportation Fuels

While hybrid-electric technology, ethanol, and hydrogen power are currently used as alternatives to oil-fueled transportation, they only comprise a small portion of total transportation. There is disagreement over whether technological improvements will allow them to replace oil on a large scale.

Environment Impact

Fossil fuels contribute to environmental degradation and public health problems, but alternatives also have environmental impacts. Wind farms, for example, kill birds and bats.

Economic Impact

Critics disagree over how increased use of energy alternatives will affect the economy. Some people believe it will provide thousand of jobs; others point out that substantial investment will be needed to develop alternatives.

Overview

Society cannot exist without energy. Throughout history, humans have relied on it to survive and to flourish. In a 2004 assessment of world energy supplies, the United Nations and the World Energy Council found, "Access to affordable energy services is fundamental to human activities, development, and economic growth."[1] Most of the energy consumed today comes from fossil fuels—primarily oil, coal, and natural gas—however, there are alternatives to fossil fuels, such as wind power, solar power, and biofuels. Because energy is such a central part of society, there is widespread disagreement about where it should come from. Policy makers, energy experts, and the general public disagree about whether alternative energy sources are necessary, what alternative sources should be pursued, whether alternative energy can be used for transportation, and how increased use of alternative energy will impact society.

Why People Need Energy

Energy is a vital part of human life and an important agent for social change. People use it to provide heat, to cool and light their homes, to refrigerate and cook their food, and to fuel transportation. It also provides power for telecommunications and commercial and industrial activity. Every year the world uses increasing amounts of energy for such purposes. However, in addition to sustaining human life, energy spurs economic and social change. For example, the use of coal allowed society to replace manual labor with machinery powered by coal. This spurred the Industrial Revolution, a major turning point in world history. Oil has also been the cause of significant social change. In the opinion of J. Andrew Hoerner of the Center for a Sustainable Economy and James Barrett of the Economic Policy Institute, "The great accomplishments of our nation in the century just ended, the enormous increase in the living standards we've enjoyed over the past one hundred years, all would have been unthinkable without the vast energy and transmission system [we have]."[2]

> **Energy is a vital part of human life and an important agent for social change.**

As energy expert Howard Geller points out, because energy is so important, those people who provide and control it have enormous power in society. He says, "Energy producers such as the major oil companies are among the largest and most profitable corporations in the world. The actions of these companies affect governments and the world economy."[3]

World Energy Consumption

The world consumes enormous amounts of energy, and the total increases every year. Energy consumption is commonly measured in British thermal units (Btu). According to the Energy Information Administration (EIA), in 1950 consumption was 229 million Btu per person in the United States. In 2005 it was 337. The EIA projects that world energy consumption will increase by 71 percent from 2003 to 2030. The United States, China, and Russia are the biggest energy consumers. The United States in particular consumes a large share of the world total—22.5 percent in

This oil pump is extracting oil from the ground. Most of the energy consumed by society currently comes from fossil fuels, of which oil is the most widely used.

2004, according to the EIA—while it contains only 4.6 percent of the world's population.

Most of the energy consumed currently comes from fossil fuels, hydrocarbons that formed in the ground over millions of years from the remains of plants and animals. The most widely used fossil fuel is oil. Oil has become an extremely important energy source especially in the United States, which consumes about a quarter of the world's oil. According to the National Resources Defense Council, in 2004 America spent approximately $270 billion to fulfill its oil needs. Says scientist Jeremy Leggett, "We have allowed oil to become vital to virtually everything we do. Ninety percent of all our transportation . . . is fueled by oil. Ninety-five percent of all goods in shops involve the use of oil. Ninety-five percent of all our food products require oil use."[4]

The Need for Alternatives

Some people argue that society cannot continue to rely on fossil fuels and must look for alternatives. One reason for this position is the widespread belief that fossil fuels are a finite resource and will eventually run out. Others insist that fossil fuels harm the environment. For example, the burning of fossil fuels is a major contributor to greenhouse gases, and the Sierra Club warns, "Unless we slow, and ultimately reverse, the buildup of greenhouse gases in the atmosphere, we will have . . . to deal with radical changes in weather patterns, sea levels and threats to human health."[5] In the opinion of the National Resources Defense Council, society must begin to reduce its oil consumption and use alternative forms of energy. It warns that in addition to the environment, oil dependence threatens national security and the economy.

> Most of the energy consumed currently comes from fossil fuels.

Others insist that because of the many advantages of fossil fuels, society will continue to use them for many years. According to the World Coal Institute, the world's increasing energy needs will primarily be met by fossil fuels. It says, "Global energy demand is projected to grow almost 60% by 2030. . . . Fossil fuels will account for the bulk of this increase and will continue to dominate the total demand for energy for the foreseeable future."[6]

Fossil fuel advocates believe that with improving technology, society will continue to discover new ways to extract fossil fuels from the earth, and will learn how to reduce environmental damage from using these fuels. Says economics professor M.A. Adelman, "It is commonly asked, when will the world's supply of oil be exhausted? The best one-word answer: Never." He says, "Growing knowledge lowers costs, unlocks new deposits in existing areas, and opens new areas for discovery."[7]

Shifting Energy Sources

History shows that in the past, society has shifted from one energy source to another. In early societies people used solar energy to heat and light their homes and wind power to grind wheat and pump water. Energy was also generated from burning wood, charcoal, and agricultural residues. Then in the 18th century coal became the main energy source. Coal powered various machines and was used to generate electricity. After World War II petroleum production increased and quickly became the dominant energy source. Use of natural gas and nuclear power has also grown in recent years. While in many developing nations people still rely on firewood and other traditional sources of energy, today wealthier nations rely on large amounts of oil and other fossil fuels to provide their energy.

> **Energy transitions have a major impact on society.**

Energy transitions have a major impact on society. For example, coal accelerated industrial development, and oil transformed mobility in transportation. The Energy Information Administration points out that transitions in energy sources have altered the way people work and where they live:

> In the middle of the 19th century, most Americans lived in the countryside and worked on farms. The country ran mainly on wood fuel and was relatively unimportant in global affairs. A hundred years later, after the Nation had become the world's largest producer and consumer of fossil fuels, most Americans were city-dwellers and only a relative handful were agricultural workers.[8]

Some people argue that in light of historical energy transitions, a transition away from oil is a natural progression. On its Web site, the Energy Information Administration summarizes the evolution of energy sources and concludes, "No doubt we have not seen the end of evolution in energy sources."[9]

Others believe that the transition from oil will be a step backward for society. According to an editorial on iTulip, a Web site that offers U.S. economic analyses by various experts, oil is a very efficient way to use energy and cannot be surpassed by any other existing technology. It insists, "The switch to oil alternatives will be a step backwards to an era when you had to expend a lot of energy just to get the fuel to the point where you can burn it."[10]

Alternatives to Fossil Fuels

While oil, natural gas, and coal are the major sources of energy today, there are numerous alternatives to these fossil fuels. Nuclear power, while it requires the fossil fuel uranium, is frequently considered to be an alternative energy source because it does not emit pollution. Wind energy, generated from wind turbines, is used to create electricity. Solar energy uses solar radiation to produce heat and electricity. Hydropower is yet another alternative, using moving water to produce electricity. Most hydropower is situated on rivers and lakes; however, many researchers believe that using ocean tides, waves, and currents to generate energy also has great potential. Another form of energy is geothermal energy, which uses geothermal heat from the interior of the planet to create electricity.

> " While oil, natural gas, and coal are the major sources of energy today, there are numerous alternatives to these fossil fuels. "

Also, there are alternatives to the oil used to fuel transportation. Hybrid-electric vehicles, which combine electricity and fuel, are available in many countries. Biofuels—derived from biomass such as plant matter—are widely used, particularly in the United States and Brazil. Many people also believe that hydrogen, while not generally available as a transportation fuel at present, has enormous potential with continued research.

Renewable Energy

Renewable sources of energy are those which are continuously replenished by natural processes, within relatively short periods of time. For example, solar energy is considered a renewable source because the sun rises again every morning, giving a fresh supply of energy regardless of how much it gave the day before. Renewable energy is generally less damaging to the environment and is unlikely to be subject to the rising costs of depletable resources such as oil. In addition, because renewables can often be produced locally rather than imported from other countries, they can provide energy security. According to the EIA, use of renewable energy in the United States has decreased significantly over the last 150 years;

in 1850 approximately 90 percent of America's energy was from renewable sources, while in 2004 it was only about 6 percent. The majority of that—approximately 70 percent—is used for electricity generation.

Nonrenewable energy sources have finite reserves and are not renewed within a person's lifetime. They may take millions of years to form, so in practical terms, once they are used they are gone forever. The main advantages of fossil fuels are that they are highly concentrated sources of energy and can be easily stored. Most of America's energy comes from nonrenewable energy sources.

Energy for Transportation

An important part of any discussion of energy alternatives is transportation, since that is what the majority of the world's oil is used for. For example, approximately two-thirds of America's oil is used for transportation. Critics point out that many commonly discussed energy alternatives such as solar, nuclear, and wind power cannot solve the world's energy problems because they are used to generate electricity, not fuel transportation. Says Jerry Taylor, a senior fellow at the Cato Institute, what is really needed are alternatives to oil. "Unfortunately," he says, "there is nothing on the horizon that comes close to gasoline as far as cost and performance is concerned."[11]

> **The main advantages of fossil fuels are that they are highly concentrated sources of energy and can be easily stored.**

Renewables advocates insist that because of the negative impacts of all the oil used for transportation, society must pursue the available alternatives, even if they are less effective than oil. For example, the use of oil for transportation also has a tremendous impact on the environment. Says the National Resources Defense Council, "Transportation . . . spews one-third of our annual emissions of heat trapping carbon dioxide pollution. In fact, one study found that the fleet of each of Detroit's automakers—GM, Ford, and DaimlerChrysler—creates more global warming pollution every year than the country's largest electric utility."[12]

Alternative Energy Use in the United States

Many people believe the United States relies too much on fossil fuels and is not doing enough to pursue alternative energy sources. According to the EIA, the United States is the world's largest energy producer, consumer, and importer. However, it does not lead the world in the use of alternative energy. The U.S. government does appear to be making some efforts to change that. In his 2007 state of the union address President George W. Bush announced his "Ten in Twenty" plan, according to which the United States will reduce its gasoline usage by 20 percent in the next 10 years. Bush claims that this will be accomplished by increasing usage of renewable fuels and increasing fuel efficiency standards. But

While oil, natural gas, and coal are the major sources of energy today, there are numerous alternatives to these fossil fuels. Wind energy, generated from wind turbines, is used to create electricity.

the plan also includes increasing domestic oil production. Critics charge that the country is still too reliant on fossil fuels and is not doing enough to develop and use alternative fuels. According to journalist Sarah Barr, the United States is far behind other nations such as Japan and Germany in developing renewable energy. "Across the world, investments in renewable energy have nearly doubled [between 2003 and 2006]," she says. "The United States is only slowly catching up with these trends."[13] The National Resources Defense Council agrees, stating, "America still depends on energy technologies of the past."[14]

Alternative Energy Use in Other Countries

Other countries are making significant and increasing use of alternatives to fossil fuels. One fast-growing form of energy is wind power. According to *USA Today* magazine, Europe produces two-thirds of the global total of wind-generated electricity. Germany, the country with the biggest capacity for generating wind energy, derives 6 percent of its electricity from wind; Spain, 8 percent; and Denmark, 20 percent. Iceland, while a small nation with a population of less than 300,000, is making significant progress in utilizing alternative forms of energy. Hydropower and geothermal plants provide all of the electricity, heat, and hot water for the nation. In total, about 70 percent of the country's energy comes from renewable sources. It is also at work on developing hydrogen energy for transportation. The country currently has some hydrogen-powered buses and future plans for passenger cars. Brazil has also made progress in utilizing alternative energy. Ethanol currently provides about 40 percent of its automobile fuel, saving the country millions of dollars in oil imports. More than 70 percent of the vehicles now sold in Brazil are flex-fuel models that run on either ethanol or gasoline.

> " The United States is the world's largest energy producer, consumer, and importer. However, it does not lead the world in the use of alternative energy. "

Some people believe that the success of energy alternatives depends on government intervention in the energy market. Explains director of the Transition Institute Karl Mallon, "Renewables are entering an existing market which is already occupied by another product, namely fossil fuels."[15] Because fossil fuels are currently cheaper, the majority of consumers choose them over more expensive alternatives. Thus, advocates of government intervention believe the government should offer subsidies or other financial help so that energy alternatives can become competitive with fossil fuels.

> [Some people] believe the government should offer subsidies or other financial help so that energy alternatives can become competitive with fossil fuels.

Others argue that government intervention is unnecessary and harmful. In the opinion of journalist Dennis Behreandt, the best way to encourage alternative energy expansion is to rely on market forces alone. He says, "The market demand for energy will spur competition to invest in new infrastructure and new technologies. . . . If left unhampered by government, there is no telling what energy technologies will be achieved in the near future by innovators eager to supply the world's increasing appetite for energy."[16]

Energy and the Economy

There is disagreement over how increased use of alternative energy would impact the economy. Some critics fear that because the economy is based on fossil fuels, replacing them with alternatives might cause job losses and recession. They argue that for society to use alternative energies on a large scale would require a huge investment and a change in the entire energy infrastructure, something that is financially impossible.

Alternatives advocates believe in an opposite scenario in which alternative energies will actually revitalize the economy. In the opinion of renewable energy expert Barry J. Hanson, an increase in renewable energy would be extremely beneficial to the economy:

> A renewable energy based economy would save the U.S. $750 billion per year plus create over six million new jobs which would revitalize local communities all across

the country. A renewable energy based economy would reduce the federal trade deficit. . . . A renewable energy based economy would require LESS government, less taxation, less spending, less welfare."[17]

Energy and the Environment

Most discussions of energy alternatives include environmental considerations. There is no doubt that the use of fossil fuels harms the environment in numerous ways. In the opinion of the International Energy Agency, "The current pattern of energy supply carries the threat of severe and irreversible environmental damage."[18] Burning fossil fuels releases pollutants that harm human health and the environment. Harvesting, processing, and transporting them can also harm the environment. For example, coal mining can transform landscapes and destroy ecosystems, and oil drilling can harm wildlife. Finally, burning fossil fuels is the largest source of carbon dioxide emissions, one of the greenhouse gases that contribute to global warming. Many experts believe that global warming will have many negative consequences for society.

> There is no doubt that the use of fossil fuels harms the environment in numerous ways.

However, others insist that fossil fuels are necessary in a modern society, and that there is no reason to stop using them, because future technology will eventually minimize environmental harms. For example, the White House National Economic Council argues that the United States needs to keep using coal for electricity and should invest in clean coal technology, which can eliminate pollutants. It also advocates expanding oil drilling in Alaska, saying, "Using modern technologies and subject to the world's most stringent environmental protections, ANWR [Arctic National Wildlife Refuge] could produce as much as 1 million barrels of oil per day to help meet our future energy needs."[19]

The Future

As Thomas C. Dorr, U.S. undersecretary for rural development, points out, history shows that the energy sources used in society do change over time. "Since the beginning of the industrial age, America's energy economy has not been static," he says. "Earlier generations of Americans transitioned from animal, wind, wood, and water power to coal, oil, natural gas, and nuclear." Dorr believes that the next step for the country is a transition to alternative sources of energy. He says, "We have managed such transitions before, and we will do so again."[20]

Others disagree that such a transition is either inevitable or desirable. As the world continues to demand increasing amounts of energy, the debate over alternatives intensifies. There is heated disagreement over whether alternative energy sources are necessary, what sources should be pursued, whether alternative energy can be used for transportation, and how increased use of alternatives will impact society.

Are Alternative Energy Sources Necessary?

> 66 **The need to curb the growth in fossil-fuel energy demand . . . is more urgent than ever.** 99

—International Energy Agency, *World Energy Outlook 2006.*

> 66 **Most alternative energy sources can be viewed as 'niche' sources, but not as a complete replacement for fossil fuels. . . . Oil will be difficult—if not impossible—to replace in its entirety.** 99

—George W. Zobrist, "Alternative Energy Sources—A Quick Look."

According to the International Energy Agency (IEA), the world needs to change the way it gets its energy. Says the organization, "[The world] is facing twin energy-related threats: that of not having adequate and secure supplies of energy at affordable prices and that of environmental harm caused by consuming too much of it."[21] It is widely agreed that the world does face certain energy problems, but there is less agreement over exactly what these problems are and whether the solutions include pursuing alternative sources of energy. The debate over whether or not energy alternatives are necessary includes assessments of the state of worldwide oil resources and whether society can or should continue to rely on them, the potential of natural gas and coal to provide energy, and the health and environmental impacts of fossil fuel use.

Worldwide Oil Resources

One key argument for the development of energy alternatives is that the world's oil supply will run out in the near future. This position is based on the premise that there is only a certain amount of oil in the ground and that at some point in time society will have used it all. According to scientist Jeremy Leggett, "At current rates of use, the global tank is going to run too low to fuel the growing demand. . . . This is not a controversial statement. It is just a question of when."[22] The IEA estimates that the world's proven oil reserves can sustain current production levels for only 42 years. Geologist Colin J. Campbell warns that the impact of running out of oil will be significant. He advises, "Make some plans for what promises to be one of the greatest economic and political discontinuities of all time."[23]

> **One key argument for the development of energy alternatives is that the world's oil supply will run out in the near future.**

Other experts point out that society continues to find new technologies for discovering and extracting oil and insist this resource will not run out. Explains Robert L. Bradley Jr., president of the Institute for Energy Research, "Resources grow with improving knowledge, expanding capital, and capitalistic policies . . . that encourage market entrepreneurship. . . . Whether or not oil, gas, and coal are exploited far into the future depends not only on consumer demand but also on whether government policies will allow the ultimate resource of human ingenuity to turn the 'neutral stuff' of the earth into resources in ever-better ways."[24] Bradley and others believe that oil supply is not limited by the depletion of resources but instead by a lack of investment in advanced technologies to extract oil from the ground.

Dependence on Foreign Oil

Critics charge that the United States is dangerously dependent on foreign oil and insist that alternative energies are part of the solution to this problem. Much of the oil used in the United States—approximately 60 percent—is imported from other countries, and because it imports so much, the country is vulnerable to disruptions in supply and price increases.

According to the White House National Economic Council, "Oil supply disruptions pose a threat to our economy and national security, and that threat rises the more dependent we are on oil imports, particularly from less stable regions of the world."[25]

One particularly volatile source of oil is the Middle East. Experts believe the region has between 50 and 60 percent of the world's oil reserves. Yet this area is also subject to political instability. For example, in 1973 Arab nations refused to sell oil to the United States because of its support for Israel. This caused massive price increases and oil shortages in the United States. Because of its reliance on foreign oil, the United States often takes diplomatic or military action to protect these oil imports. Says energy expert Howard Geller, "Western nations spend tens of billions of dollars each year protecting oil supplies in the Middle East."[26] The White House Economic Council and others believe that a better solution is to reduce America's foreign oil consumption and that one way to do this is to develop domestic energy alternatives.

> **[Some people] believe that oil supply is not limited by the depletion of resources but . . . by a lack of investment in advanced technologies to extract oil.**

Difficulty of Replacing Oil

Some people believe that because of the way society has come to depend on fossil fuels, it will be extremely difficult to replace them, particularly with the currently available alternatives. Marty Bender, scientist at the Land Institute in Kansas, explains:

> We do not get as much energy back from renewable energy sources as from fossil fuels. During the past century, the U.S. economy has largely been powered by fossil fuels that returned as much as 100 times the energy spent taking them from the ground. In contrast, the ethanol in your car's gasohol barely contains more energy than was needed to make it from corn. Other renewable fuels are only slightly better. . . . This means we cannot expect re-

newables to produce the amount of energy our economy now consumes.[27]

Yet renewables advocates contend that even if it seems impossible, society must attempt to replace fossil fuels. They believe such a transition will eventually be successful. Says Leggett, "When it becomes clear that there is no escape from ever-shrinking supplies of . . . oil . . . human society will face an energy crisis of unprecedented proportions. . . . To avoid it, we need to make the mass use of alternatives to oil, gas, and coal the main principle around which we organize society and economies."[28] He insists that this is a realistic goal, saying, "In a society that put a man on the Moon more than three decades ago, surely there can be no doubt that we could replace oil if we seriously wanted to."[29]

Natural Gas

While oil reserves in the United States are quickly dwindling, the country has relatively large natural gas reserves and also many different opinions on the role this fossil fuel will play in the nation's future energy supply. According to the White House National Economic Council, in 2004 the United States produced 85 percent of its natural gas. Natural gas has many uses including heating buildings, heating water, cooking, drying clothes, lighting, and industrial purposes. Says executive director of the Progressive Policy Institute Chuck Alston:

> Demand for natural gas is growing because it is relatively cheap, it is efficient, it is the cleanest burning fossil fuel, and the United States largely meets its needs with domestic sources. These traits make it vital not only to the nation's energy future, but also to the challenge of reducing harmful emissions from power plants, including carbon dioxide (CO_2), the chief culprit in global warming.[30]

> **The United States often takes diplomatic or military action to protect [its] oil imports.**

The National Energy Technology Laboratory agrees that natural gas should remain a significant part of America's energy supply. It says,

"Our Nation has a vital interest in ensuring that competitively-priced domestic natural gas . . . remain[s] part of the U.S. energy portfolio for decades to come."[31]

However, critics caution that natural gas has the same problems as other fossil fuels. As Raymond J. Kopp, writing for *Resources for the Future* says, "Like petroleum and coal, natural gas is an exhaustible resource, and at some point the world may run short of it."[32] Critics also point out that although natural gas may be cleaner burning than oil, it is not pollution free.

The Potential of Coal

It is generally agreed that there are still vast reserves of coal around the world, and some people believe this fossil fuel will become an increasingly important energy source. The International Energy Agency estimates that proven worldwide coal reserves will last for 164 years at current production rates. The United States, Russia, and China have approximately half of these reserves. According to the White House National Economic Council, coal provides more than half of the electricity used by Americans, and that may increase in the future. David Talbot, chief correspondent for the Massachusetts Institute of Technology's *Technology Review* argues, "Coal is the most abundant fossil fuel. . . . We're stuck with coal. Since there's little reason to expect that humankind will stop digging for it, we will have to find cleaner ways to burn it."[33]

> " Renewables advocates contend that even if it seems impossible, society must attempt to replace fossil fuels. "

While many critics charge that burning coal harms the environment, coal advocates insist that clean-burning coal technology is available and improving. According to the World Coal Institute, "Significant reductions in CO_2 emissions from coal-fired power stations have already been achieved through increasing efficiency. . . . An ultra low emissions future is achievable with the development and deployment of the next generation technologies."[34]

Skeptics such as Leggett believe that increasing the use of coal will only be harmful to society's future. Leggett explains that coal has the potential to harm in many ways. He says, "Besides the future death toll from unmitigated global warming and dire air quality, there is also the actual death toll to date from getting the stuff out of the ground. Type 'coal mining disasters' into Google and see what I mean."[35] Physicist David Goodstein agrees that coal is not a good long-term energy solution because like oil, it will eventually run out. He argues, "We will run out of all fossil fuels. Coal will peak just like any natural resource. We will reach the peak for all fossil fuels by the end of the century."[36]

Coal advocates insist that clean-burning coal technology is available.

Health Threats and Environmental Degradation

One compelling reason to pursue energy alternatives is the impact that fossil fuels have on public health and on the environment. The burning of fossil fuels produces pollutants that are a major public health threat, especially to the elderly, sick, and very young. It releases mercury into the environment, which causes neurological damage in children. Fossil fuels also contribute to acid rain, ozone, and carbon dioxide pollution; production, transportation, and use of oil can cause water pollution; and coal mining can pollute water and destroy soil. According to the National Resources Defense Council, "Burning fossil fuels in our power plants, cars, and factories accounts for more than 60,000 premature deaths in the United States each year. And drilling for oil and gas industrializes some of our most prized wild places, denying future generations the last remnants of our natural heritage."[37]

Dallas Burtraw and Karen L. Palmer of Resources for the Future insist that if society decreases its use of fossil fuels, it will benefit substantially. They argue, "Reducing these pollutants . . . will yield tens of billions of dollars per year in public health benefits."[38]

The National Center for Policy Analysis is one organization that believes that the environmental harms of fossil fuels have been exaggerated. It says, "Fossil fuel use does produce a degree of environmentally harmful side effects. However, these are generally greatly overstated. Despite the fact that energy use requiring fossil fuels has increased dramatically over the last 30 years, air and water pollution, the primary environmental side-effects of fossil fuel use, has declined just as dramatically."[39] The Environmental Protection Agency agrees that there have been substantial improvements in U.S. air quality in recent years.

Global Warming

Some people believe that the biggest environmental problem caused by fossil fuels is global warming. When these fuels are burned, they release carbon dioxide and other greenhouse gases into the atmosphere. Many scientists believe this is causing Earth's temperature to rise. Leggett talks about the potentially catastrophic impacts of global warming. He says:

> Some 6 percent of all the freshwater on the planet . . . sits frozen . . . atop Greenland. It can melt. If it did so, global sea levels would rise 7 meters, inundating coastal cities and plains where most of the world's population lives, and where most economic and agricultural activity is focused. . . . In the Amazon, rainfall is forecast to drop, leading to the gradual death of the rainforest, massive biodiversity reductions, and the release of additional carbon into the atmosphere."[40]

> **Some people believe that the biggest environmental problem caused by fossil fuels is global warming.**

Critics of global warming insist there is no proof that temperature increases will harm society and that any attempt to reduce greenhouse gases will be harmful to the economy. In the opinion of commentator George F. Will, "We do not know how much we must change our economic activity to produce a particular reduction of warming. And we do not know whether warming is necessarily dangerous. Over the millennia, the planet has warmed and cooled for reasons that are unclear but clearly were unrelated to SUVs." He asks, "Are we sure that the climate at this particular moment is exactly right, and that it must be preserved, no matter the cost?"[41]

Society continues to disagree on whether alternative energy sources are necessary or practical. Karl Mallon, director of the Transition Institute, suggests that even if alternative energy sources are not a perfect alternative to fossil fuels, they should be pursued because they can help address problems with the current energy supply. He says, "Renewable energies are the best thing we have so far. . . . If they meet all our future needs in due course that is great, but if they do nothing more than buy time while better technologies are evolved that is also good."[42] Many communities have taken this approach, using various energy alternatives despite their problems. However, such use has not eliminated disagreement over the necessity of these technologies, including debate on the state of oil reserves, the potential of natural gas and coal, and the health and environmental impacts of fossil fuel use.

Are Alternative Energy Sources Necessary?

66 **When it becomes clear that there is no escape from ever-shrinking supplies of . . . oil . . . society will face an energy crisis of unprecedented proportions. . . . To avoid it, we will need to make . . . use of the alternatives to oil, gas, and coal.** 99

—Jeremy Leggett, *The Empty Tank.* New York: Random House, 2005, pp. xiv–xv.

Leggett is an award-winning scientist, oil industry consultant, Greenpeace campaigner, and author *of Half Gone: Oil, Gas, Hot Air, and the Global Energy Crisis.*

66 **An adequate and affordable supply of energy and resources is a key ingredient in a prosperous U.S. economy, and oil remains one vital component of our energy portfolio.** 99

—Ian Parry and Joel Darmstadter, "Slaking Our Thirst for Oil," in *New Approaches on Energy and the Environment: Policy Advice for the President,* eds. Richard D. Morgenstern and Paul R. Portney. Washington, DC: Resources for the Future, 2004, p. 27.

Parry and Darmstadter are senior fellows at Resources for the Future, a nonprofit organization that conducts research on environmental, energy, and natural resource issues.

Bracketed quotes indicate conflicting positions.

* Editor's Note: While the definition of a primary source can be narrowly or broadly defined, for the purposes of Compact Research, a primary source consists of: 1) results of original research presented by an organization or researcher; 2) eyewitness accounts of events, personal experience, or work experience; 3) first-person editorials offering pundits' opinions; 4) government officials presenting political plans and/or policies; 5) representatives of organizations presenting testimony or policy.

66As apparent as it seems to many that the nation should do 'something' about Energy, the . . . complexity [of the issue] augurs against easy agreement to . . . the policy options.99

—Robert Bamberger, "Energy Policy: Conceptual Framework and Continuing Issues," Congressional Research Service, December 6, 2006. www.ncseonline.org.

Bamberger is a specialist in energy policy at the Resources, Science, and Industry Division of the Congressional Research Service.

66Our Nation's reliance on oil leaves us vulnerable to hostile regimes and terrorists. . . . We need to diversify our energy supply.99

—George W. Bush, radio address, February 10, 2007. www.whitehouse.gov.

Bush is the forty-third president of the United States.

66A lot of oil comes from unstable parts of the world. Producer instability, however, is not necessarily a good reason to abandon oil.99

—Jerry Taylor and Peter Van Doren, "Stuck on Empty," *National Review*, February 3, 2006. www.nationalreview.com.

Taylor and Van Doren are senior fellows at the Cato Institute in Washington, D.C. Van Doren is editor of Cato's *Regulation* magazine.

66Using portions of the hundreds of billions of petrodollars they are annually draining from our economy, Middle Easterners have established training centers for terrorists. . . . We must take ourselves . . . off the petroleum standard.99

—Robert Zubrin, "An Energy Revolution," *American Enterprise,* March 2006. www.taemag.com.

Zubrin is president of the aerospace engineering and research firm Pioneer Astronautics and author of *The Case for Mars* and other books.

> **66Natural gas remains [one of] the primary fuel sources to meet American energy needs.99**

—Independent Petroleum Association of America, *2007 Oil and Natural Gas Issues Briefing Book,* 2007. www.ipaa.org.

The Independent Petroleum Association of America is a national association that represents independent oil and natural gas producers in the United States.

> **66Natural gas is not sufficiently clean to be considered the long-term answer to America's energy needs.99**

—National Resources Defense Council, "A Responsible Energy Plan for America," April 2005. www.nrdc.org.

The National Resources Defense Council is a nonprofit organization of lawyers, scientists, and environmentalists dedicated to protecting the environment and public health.

> **66Coal has a unique role to play in meeting the demand for a secure energy supply. Coal is . . . a reliable, secure and affordable fuel.99**

—World Coal Institute, "Coal: Secure Energy," October 2005. www.worldcoal.org.

The World Coal Institute is a nonprofit, nongovernmental association that represents the coal industry in international energy and environmental policy and research discussions.

> **66New conventional coal plants in the age of global warming are not just bad policy—they are a bad investment, and one we cannot afford to make.99**

—Barbara Freese and Steve Clemmer, "Gambling with Coal: How Future Climate Laws Will Make New Coal Power Plants More Expensive," Union of Concerned Scientists, September 2006. www.ucsusa.org.

Freese is an attorney and consultant and Clemmer is a research director for the Union of Concerned Scientists.

❝[The world needs to] substantially increase with a sense of urgency the global share of renewable energy in the total energy supply.❞

> —Bonn Renewables Energy Conference 2004, political declaration, June 4, 2004.

The Bonn Renewables Energy Conference hosted by Germany in 2004 was organized in order to encourage the expansion of renewable energies worldwide.

❝Long-term solutions to enable reliable, affordable energy must include massive investment into finding new supplies of oil and gas.❞

> —ExxonMobil, "Responding to World Energy Needs," January 2007. www.exxonmobil.com.

ExxonMobil is an oil and gas company and one of the largest and wealthiest companies in the world.

❝Only by a Manhattan Project–scale government effort to develop green energy can we hope to avert the worst consequences of global warming.❞

> —Juan Cole, "Al Gore, Global Warming, the Oscars, and the Iraq War," *Informed Comment*, February 26, 2007. www.juancole.com.

Cole is a professor of modern Middle East and South Asian history at the University of Michigan.

❝The most inconvenient truth about global warming is that we cannot stop it. . . . Even if the world adopted the most far-reaching plans to combat climate change.❞

> —Fareed Zakaria, "Unfortunately, We Can't Stop Global Warming," *Newsweek*, February 19, 2007. www.msnbc.msn.com.

Zakaria is the editor of *Newsweek International*. He also writes a regular column for *Newsweek* and is an analyst for *ABC 7 News*.

Facts and Illustrations

Are Alternative Energy Sources Necessary?

- The White House National Economic Council predicts that by **2030** the United States will be importing **62.5 percent** of its total oil.

- According to the National Resources Defense Council, the world uses about **12 billion** more barrels of oil per year than it finds.

- Saudi Arabia has the largest known oil reserves in the world, and in 2004 Saudi minister of petroleum and mineral resources Ali Al-Naimi reported that there will be **no shortage** of oil for at least the next **50 years**.

- The National Resources Defense Council estimates that during peacetime the United States spends **$20 to $40 billion** per year in military costs to secure access to foreign oil supplies.

- According to the Energy Information Administration, approximately **23 percent** of America's energy consumption is supplied by natural gas.

- The United States produces about **85 percent** of the natural gas that it uses.

- According to the World Coal Institute, coal generates **40 percent** of the world's electricity.

- The White House National Economic Council reports that the United States has more than **25 percent** of the world's coal reserves.

Persian Gulf Largest Source of U.S. Oil Imports

According to this graph of U.S. crude oil imports, the biggest source of imported oil is the Persian Gulf. Mexico, Venezuela, and Canada also supply a significant proportion of America's imported oil.

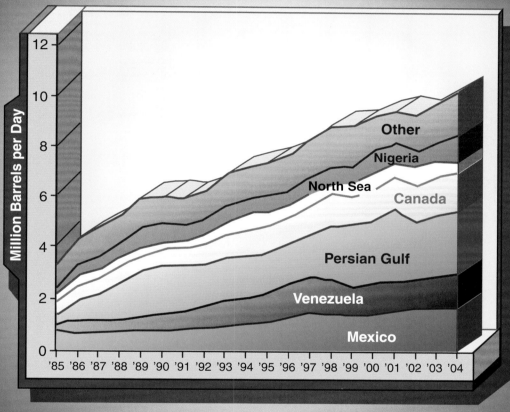

Source: Energy Information Administration, "United States: Oil," *Country Analysis Briefs*, November 2005. www.eia.doe.gov.

- According to the Massachusetts Institute of Technology *Technology Review*, coal consumption produces **37 percent** of world emissions of carbon dioxide, the primary greenhouse gas.

World Coal Reserves Exceed Oil and Gas

This map reveals that the world's overall proven coal reserves exceed those of conventional oil and gas. Only South America and the Middle East have greater oil reserves than they do coal.

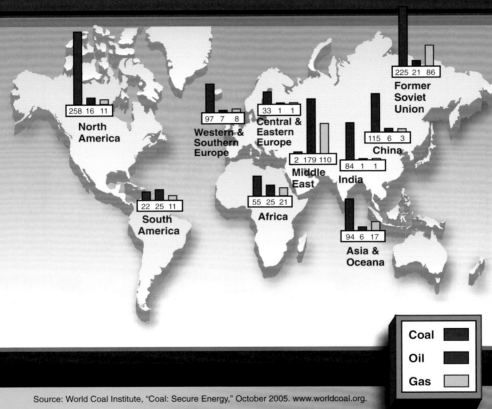

| 258 16 11 |
| North America |

| 97 7 8 |
| Western & Southern Europe |

| 33 1 1 |
| Central & Eastern Europe |

| 225 21 86 |
| Former Soviet Union |

| 115 6 3 |
| China |

| 2 179 110 |
| Middle East |

| 84 1 1 |
| India |

| 22 25 11 |
| South America |

| 55 25 21 |
| Africa |

| 94 6 17 |
| Asia & Oceana |

Coal
Oil
Gas

Source: World Coal Institute, "Coal: Secure Energy," October 2005. www.worldcoal.org.

- According to the **Central Intelligence Agency**, the United States is the largest single emitter of carbon dioxide from the burning of fossil fuels.
- The Foundation for Clean Air Progress reports that since **1978** air pollution in the United States has **decreased** dramatically.

Fossil Fuels Contribute Greatly to Environmental Harm

This chart shows the cause of some common environmental harms. It reveals that the burning of fossil fuels for commercial energy supply and manufacturing are a major cause of many pollutants.

Share of human disruption caused by

Pollutant	Commercial energy supply	Traditional energy supply	Agriculture	Manufacturing, other
Lead emissions to atmosphere	41% (fossil fuel burning, including additives)	Negligible	Negligible	59% (metals processing, manufacturing, refuse burning)
Oil added to oceans	44% (petroleum, harvesting, processing, transport)	Negligible	Negligible	56% (disposal of oil wastes, including motor oil changes)
Mercury emissions to atmosphere	20% (fossil fuel burning)	1% (burning traditional fuels)	2% (agricultural burning)	77% (metals processing, manufacturing, refuse burning)
Particulate emissions to atmosphere	20% (fossil fuel burning)	10% (burning traditional fuels)	40% (agricultural burning)	15% (smelting, non-agricultural land clearing, refuse)
CO_2 flows to atmosphere	75% (fossil fuel burning)	3% (net deforestation for fuel wood)	15% (net deforestation for land clearing)	7% (net deforestation for lumber, cement, manufacturing)
Sulfur emissions to atmosphere	85% (fossil fuel burning)	0.5% (burning traditional fuels)	1% (agricultural burning)	13% (smelting, refuse burning)

Source: United Nations Development Program, United Nations Department of Economic and Social Affairs, and the World Energy Council, "World Energy Assessment: Overview–2004 Update," 2004. www.undp.org.

Carbon Dioxide Levels Are Increasing

According to this chart, the world's carbon dioxide levels are currently higher than they have been at any time over the past 650,000 years. Carbon dioxide is a primary cause of global warming.

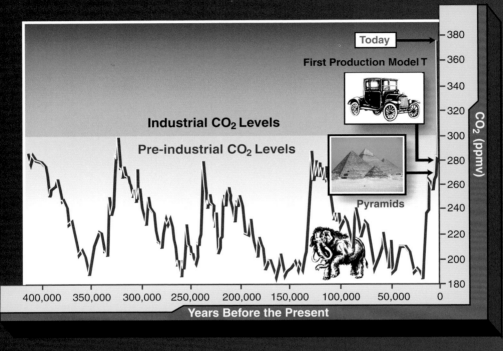

Source: Union of Concerned Scientists, "Global Warming," Novermber 8, 2006. www.ucsusa.org.

- The U.S. Department of Energy estimates that approximately **50 percent** of Americans live in areas where levels of one or more air pollutants are high enough to affect public health and/or the environment.

- In a 2006 *Time* magazine poll of **1,002 adult Americans**, 85 percent said they believe global warming is occurring.

- The Intergovernmental Panel on Climate Change predicts that global warming will cause world temperatures to rise between **3.2 and 7.2 degrees** Fahrenheit by the end of the century.

What Alternative Energy Sources Should Be Pursued?

> 66 All of . . . [alternative energy] technologies have some problems. 99

—Mat Conway, "Alternative Energy—Things You Didn't Know."

> 66 No single solution can meet our society's future energy needs. The answer lies instead in a family of diverse energy technologies that share a common thread: they do not deplete our natural resources or destroy our environment. 99

—Union of Concerned Scientists, "Clean Energy."

In his book about energy alternatives, renewable energy expert Barry J. Hanson enthusiastically supports the transition from fossil fuels to an alternative energy society. He writes, "Exciting new breakthrough technologies make transitioning to a renewable energy economy a practical reality unforeseen even a few short years ago."[43] However, the reality is that alternative energy still plays a relatively minor role in the overall energy supply. One primary reason for this is the many difficulties encountered in the installation and use of these different energy sources. Disagreement continues on whether nuclear, solar, wind, hydropower, geothermal, and ocean energy can be used to supply large amounts of energy for society and what the advantages and disadvantages of each technology are.

Nuclear Power

Some people believe that for electricity production, nuclear power is an effective alternative to fossil fuels. They point out that nuclear power plants do not pollute the air and are relatively inexpensive to run after the initial cost of construction has been paid for. According to the White House National Economic Council, nuclear power is the second largest source of electricity generation in the United States, providing about one-fifth of the country's electricity. The United States has over 100 operating nuclear plants. *National Geographic* writer Michael Parfit reports that 440 nuclear plants worldwide generate 16 percent of the world's electricity. France derives 78 percent of its electricity from nuclear power. In the opinion of Patrick Moore, one of the founders of Greenpeace, "[There are] enormous and obvious benefits of harnessing nuclear power to meet and secure America's growing energy needs. These benefits far outweigh any risks."[44]

Moore says that nuclear power is an excellent way to meet growing energy demands while reducing greenhouse emissions and believes America needs to invest in developing more nuclear power. He says, "Technology has now progressed to the point where the fear-mongering being spread by activists about the safety of nuclear energy bears no semblance to reality."[45]

> **Alternative energy still plays a relatively minor role in the overall energy supply.**

Critics of nuclear power point out some problems with this energy source. First, nuclear power plants require a substantial initial investment; between $2 billion and $3.5 billion per reactor, according to the International Energy Agency. The planning and construction of nuclear plants also takes a long time. In addition, say critics, after the plant is completed, the possibility of accidents and the difficulty of disposing of radioactive waste still remain. Fears of a nuclear accident often evoke memories of the 1986 Chernobyl accident that occurred in the former Soviet Union, the worst in the history of nuclear power. This disaster contaminated large areas, resulting in the evacuation of many people and long-lasting health problems for many more. According to a 2005 report by the Chernobyl Forum, led by the International Atomic Energy Agency and the World Health Organization, as many as 9,000 people may die from some form of cancer as a result of exposure to contamination from that accident. Critics also charge that nuclear power is not renewable because it requires uranium. The IEA estimates that under current usage rates, known uranium reserves will last for only several decades.

> " **Nuclear power plants require a substantial initial investment; between \$2 billion and \$3.5 billion per reactor.** "

Solar Power

Solar power converts the sun's energy into heat, lighting, and electricity. People continue to disagree over its potential. Advocates like journalist Bill McKibben insist that the sun's energy offers great potential that is just waiting to be used. He says, "If the sun's out, it's hitting your roof right now, and bouncing back unused into the atmosphere . . . a gift refused."[46]

However, others remain skeptical, arguing that solar power has too many problems to become a major source of energy. Responding to a *Reason* magazine article about the potential of solar power, reader James contends, "Solar power has been 'just around the corner' for my whole lifetime. Among other problems? The sun doesn't shine at night, but people still need electricity. . . . What about cloudy days?"[47]

Solar power advocates insist that it is simply a matter of additional research and development to improve solar technology. Physicist David Goodstein argues, "If you want to gather enough solar energy to replace the fossil fuel that we're burning today . . . [it] is not impossible. It's just difficult. It's hard and we're not trying."[48]

Cost and Efficiency of Solar Power

A primary critique of solar power is that initial installation is too expensive for the average person to afford and that it takes too long for the investment to pay off. While advocates such as the National Resources

Defense Council point out that the cost of solar has decreased and predict that it will continue to do so, others such as journalist Kimberly Lisagor insist that it is still very expensive. When considering solar energy for her own home, Lisagor finds that, "Solar energy would require an up-front investment that could put a sizeable dent in my budget."[49] Specifically, says Lisagor, installing solar panels on her 1,400 square-foot California home would cost her more than $20,000, which at current electricity rates will take 29 years to pay off.

> " Solar power advocates insist that it is simply a matter of additional research and development to improve solar technology. "

Critics also maintain that solar power is inefficient and disagree with proponents over whether that will limit its widespread use. According to Jay Lehr, science director for the Heartland Institute, "A short description of the solar problem is that no matter how you design the system it will always be inefficient and capture only a small, uneconomical amount of solar energy. The best solar cells available on a large scale have an efficiency of about 10 percent—they can only capture about 10 percent of the solar energy that strikes the cells."[50]

He believes solar power will never be able to provide a significant percentage of energy. U.S. secretary of energy Samuel Bodman contends that efficiency is increasing with additional research and that in only a few years the United States will be able to power as many as 2 million more homes with solar energy.

Wind Power

Wind power entails the conversion of wind energy into a form society can use—usually electricity—using wind turbines. This alternative form of energy is expanding rapidly in both the United States and around the world. According to the White House National Economic Council, "Wind energy is one of the world's fastest-growing energy technologies."[51] The National Resources Defense Council finds that wind energy in the United States is expanding at a rate of more than 20 percent a year. It has grown even faster in Europe, which leads the world in wind power.

Spain and Germany generate large amounts of wind power, and in Denmark it supplies about 20 percent of the country's electricity needs.

Proponents of wind technology maintain that it has many benefits over other forms of electricity generation. Wind farms can be built much more quickly than coal or natural gas plants and recoup their investment very quickly. According to *USA Today* magazine, since the 1980s the cost of wind power has decreased by almost 90 percent. Wind advocates believe that wind power is much better for the environment because wind farms do not pollute. Finally, wind power enhances energy security because it can be produced locally and is inexhaustible.

However, critics point out that wind power is not perfect. Wind farms require large areas of land and can be unsightly; for this reason numerous communities do not want wind farms in their cities. Another problem is that when the wind stops blowing, consumers must turn to another source of energy such as coal-fired power plants. In summary, according to H. Sterling Burnett, senior fellow at the National Center for Policy Analysis, "Renewable energy promoters claim that wind power is cheap, safe and 'green.' These claims are untrue."[52]

Wind Power's Impact on Birds and Bats

One of the primary critiques of wind farms is their impact on birds and bats. Critics charge that wind turbines can have a devastating impact on these populations. According to journalist Michelle Nijhuis, "It's well known that wind turbines kill both birds and bats."[53] The worst case documented is at the Altamont wind farm outside San Francisco, where an estimated 1,300 golden eagles, hawks, and other raptors are killed every year. Defenders of wind power insist that Altamont is an anomaly. They say that the industry is doing a much better job of choosing sites that will be less harmful to birds and is using technology in ways that minimize harm to the birds. The National Resources Defense Council finds that, "Project developers generally work with local bird experts to avoid migration routes and new technologies are being used to help birds steer clear of turbines."[54] Tom Gray of the American Wind Energy Asso-

> " Wind turbines can have a devastating impact [on birds and bats]. "

ciation concludes, "A number of studies at other projects indicate that, in general, wind power is not a significant threat to birds."[55]

However, others caution that not enough research is available at present to determine how wind farms are really affecting birds and bats. Says David Klute of the Colorado Division of Wildlife, "The technology is so new, and the potential impacts are so new, that a lot of our biologists just haven't dealt with them."[56]

Hydropower

Hydropower is the capture of energy from moving water. Today, most hydropower is generated with water flowing from a higher level to a lower level, such as over a dam. According to the EIA, of total U.S. electricity generation from renewable energy in 2004, hydropower accounted for 75 percent. The agency says that over one-half of total U.S. hydroelectric capacity for electricity generation is concentrated in Washington, California, and Oregon. Says the EIA, "Some people regard hydropower as the ideal fuel for electricity generation because, unlike the nonrenewable fuels used to generate electricity, it is almost free, there are no waste products, and hydropower does not pollute the water or the air."[57] Another advantage of hydropower is its flexibility in responding to changing electricity demands and its storage capacity. The U.S. Hydropower Council for International Development believes that hydropower has enormous potential in the United States and elsewhere. It estimates that only one-third of the world's hydropower potential has been tapped.

> **Of total U.S. electricity generation from renewable energy in 2004, hydropower accounted for 75 percent.**

However, hydropower is not environmentally benign. The major criticism is that it alters natural habitats. Hydropower often requires the creation of dams, which can change the flow of rivers, affecting the animals and the people that depend on that water. For example, the Environmental Protection Agency explains,

> Certain salmon populations in the Northwest depend on rivers for their life cycles. These populations have been

dramatically reduced by the network of large dams in the Columbia River Basin. When young salmon travel downstream toward the ocean, they may be killed by turbine blades at hydropower plants. When adult salmon attempt to swim upstream to reproduce, they may not be able to get past the dams."[58]

Geothermal Energy

Geothermal energy is generated using heat from the earth. Many people believe it is a promising source of energy for the future. In 2006 the Massachusetts Institute of Technology (MIT) evaluated the potential for geothermal energy as an energy source in the United States. In its concluding report, MIT states that many attributes of geothermal energy make it desirable: low emissions, a small environmental footprint, and widespread availability across the country. It can also be extracted without burning fossil fuels. In addition, as the U.S. Department of Energy points out, "Geothermal energy is *available 24 hours a day,* 365 days a year."[59] At present, most U.S. hydrothermal generation occurs in the western states, Alaska, and Hawaii, but the Department of Energy believes that significant expansion is possible.

> " Geothermal energy has the potential to harm the environment. "

Yet, geothermal energy has the potential to harm the environment. In an assessment of the environmental impact of this alternative energy form, researchers Karl Gawell and Diana Bates found, "Environmental effects of withdrawing steam from the earth are several. Air pollution, thermal pollution, water pollution, land subsidence, groundwater contamination, and possible earth tremors constitute a formidable array of unanswered environmental questions."[60]

Ocean Energy

Ocean energy uses the power of ocean currents, waves, and tides to create energy. Its advocates believe it has enormous potential as a future energy source. As journalist Eric Scigliano explains, "The tsunami in the Indian Ocean last December [2004] that killed nearly 300,000 people . . . offered the world an indelible demonstration of how much energy a wave can carry." He says, "The bonanza is so obvious that inventors have dreamed of harnessing ocean waves for more than two centuries."[61]

Some potential advantages of ocean power are that the ocean is constant and predictable and contains a huge amount of power. However, of all the energy alternatives this is one of the least developed. Problems include the very harsh environment of the sea. Ocean water corrodes and storms damage or displace equipment. In addition, many people are concerned that the equipment used to generate power will be unsightly and mar scenic landscapes. However, there is widespread optimism about using ocean energy. The Massachusetts Technology Collaborative says, "While very few of these technologies have been implemented on a commercial scale, they show much promise for future development."[62]

> **Advocates [of ocean energy] believe it has enormous potential as a future energy source.**

Because each energy alternative has advantages and disadvantages, some people believe there is no single answer to the question of what energy sources to pursue. According to *National Geographic* contributor Michael Parfit, in order to replace fossil fuels, humans will have to embrace many alternatives. He says, "Plenty of contenders for the energy crown now held by fossil fuels are already at hand: wind, solar, even nuclear, to name a few. But the successor will have to be a congress, not a king."[63] Critics continue to disagree on the relative benefits of all these contenders though, debating the potential for nuclear, solar, wind, hydropower, geothermal, and ocean energy to provide energy for society.

What Alternative Energy Sources Should Be Pursued?

> 66 Safe nuclear power is a critical part of meeting our future energy needs. . . . It is the only mature power generating source that can meet our current and future needs with zero emissions. 99

—Samuel Bodman, "Ask the White House," October 2, 2006. www.whitehouse.gov.

Bodman is the U.S. secretary of energy.

> 66 Nuclear power . . . remains riddled with political difficulties both with regard to location and to the handling of nuclear waste. Moreover, over the whole scene hangs the growing terrorist risk—poised . . . to inflict deadly damage. 99

—David Gore-Booth, "An Overview," in *Emerging Threats to Energy Security and Stability,* eds. Hugo McPherson, W. Duncan Wood, and Derek M. Robinson. Netherlands: Springer, 2005, p. 16.

Gore-Booth served as a British diplomat in the Middle East. He died in 2004.

Bracketed quotes indicate conflicting positions.

* Editor's Note: While the definition of a primary source can be narrowly or broadly defined, for the purposes of Compact Research, a primary source consists of: 1) results of original research presented by an organization or researcher; 2) eyewitness accounts of events, personal experience, or work experience; 3) first-person editorials offering pundits' opinions; 4) government officials presenting political plans and/or policies; 5) representatives of organizations presenting testimony or policy.

❝My experience in the nuclear industry dates back 36 years. . . . [I believe] nuclear energy must be part of America's energy future.❞

—Anthony F. Earley Jr., "The Nuclear Renaissance: Is It Real?" speech to the Economic Club of Detroit, February 12, 2007. www.dteenergy.com.

Earley is energy chairman and chief executive officer of energy company DTE Energy.

❝The greatest potential for . . . [large]-scale renewable electric power lies in harvesting solar energy.❞

—Marty Hoffert, "It's Not Too Early," *Technology Review*, July 17, 2006. www.technologyreview.com.

Hoffert is professor emeritus of physics at New York University.

❝For decades, there have been delirious proclamations that the world would soon run on solar energy. Those statements . . . always have been false.❞

—Jay Lehr, "Solar Power: Too Good to Be True," *Heartland Institute*, June 1, 2005. www.heartland.org.

Lehr is science director at the Heartland Institute, a nonprofit organization that promotes free market solutions to social and economic problems.

❝Julie and I have lived in a solar-powered house for seven years. . . . [Many people thought] we were holed up in a shack. . . . The reality . . . [is] that we were watching DVDs and Satellite TV on our home theater system and munching microwave popcorn while the dishwasher hummed away in the background.❞

—Ric Soulen and Julie Soulen, "A Solar Power Primer," *Colorado Journal*, February 22, 2007. www.denverpostbloghouse.com.

The Soulens live in a solar-powered house in Colorado.

66Wind power is an affordable, efficient and inexhaustible source of electricity. It is pollution-free, and thanks to technological breakthroughs, it's cost competitive.99

—National Resources Defense Council, "Wind, Solar, and Biomass Energy Today," January 12, 2006. www.nrdc.org.

The National Resources Defense Council is a nonprofit organization of lawyers, scientists, and environmentalists dedicated to protecting the environment and public health.

66Wind power is expensive, doesn't deliver the environmental benefits it promises and imposes substantial environmental costs.99

—H. Sterling Burnett, "Wind Power: Red, Not Green," *Brief Analysis No. 467*, National Center for Policy Analysis, February 23, 2004. www.ncpa.org.

Burnett is a senior fellow with the National Center for Policy Analysis, an organization that works to develop and promote private alternatives to government regulation.

66I would love to live next to a wind farm, I really would. My parents in Scotland are just in the process of buying a new house and my mother was quite upset when they had to sell the old one because she had a beautiful view of a wind farm from her kitchen window.99

—Alison Hill, "Renewable Energy," interview with Alison Hill and Max Carcas, *Naked Scientists,* January 2007. www.thenakedscientists.com.

Hill is communications manager for the British Wind Energy Association.

❝Hydropower plants produce no air pollutants. . . . Moreover, hydropower projects do not generate any toxic by-products.❞

> —International Hydropower Association, "Top Ten Reasons Why Hydropower Should Lead the Way in Clean Development Mechanism," 2004. www.hydropower.org.

The International Hydropower Association is a nongovernmental association with members in more than 80 countries. It aims to advance hydropower's role in meeting the world's energy needs.

❝Although hydropower has no air quality impacts, construction and operation of hydropower dams can significantly affect natural river systems as well as fish and wildlife populations.❞

> —Environmental Protection Agency, "Electricity from Hydropower," July 19, 2006. www.epa.gov.

The Environmental Protection Agency is an agency of the federal government charged with protecting human health and with safeguarding the natural environment.

❝The exciting field of wave and tidal-delivered energy . . . is already providing steady and cost-effective energy in certain locales. . . . Let's ride the tide of wave energy and watch our electricity bills decline.❞

> —Cynthia Thielen, "Hawaii Should Ride the Tide of Wave Energy," *Star Bulletin*, October 3, 2005. http://starbulletin.com.

Thielen is a state representative in Hawaii and the assistant minority floor leader.

❝Geothermal energy . . . can provide . . . electric power and heat at a level that can have a major impact on the United States, while incurring minimal environmental impacts.❞

—Massachusetts Institute of Technology, "The Future of Geothermal Energy," 2006. http://geothermal.inel.gov.

The Massachusetts Institute of Technology is a private research university located in Cambridge, Massachusetts.

❝Environmental effects of withdrawing steam from the earth are several. Air pollution, thermal pollution, water pollution, land subsidence, groundwater contamination, and possible earth tremors. . . . It suffices to say that geothermal power is not environmentally free.❞

—Karl Gawell and Diana Bates, "Geothermal Literature Assessment: Environmental Issues," May 2004. www.geothermal-biz.com.

Gawell is executive director of the Geothermal Energy Association. Bates was a former researcher for the association.

What Alternative Energy Sources Should Be Pursued?

- The Energy Information Administration estimates that in 2005, **9 percent** of America's total energy production and **6 percent** of its consumption was from renewable sources.

- According to the International Energy Agency, in 2005 **nuclear power plants** supplied **15 percent** of the world's electricity.

- Nuclear power provides approximately **20 percent** of the United States' electricity.

- According to a report cited by the U.S. Department of Energy, in 2006 worldwide installation of solar photovoltaic devices increased by **19 percent** over 2005. Germany accounted for **55 percent** of that increase.

- The Energy Information Administration finds that about **90 percent** of America's solar energy is used for heat.

- According to *USA Today* magazine, in 2005 global wind electricity-generating capacity increased **24 percent**.

- *USA Today* magazine finds that since the 1980s the cost of wind power has decreased by almost **90 percent**.

- The White House National Economic Council predicts that areas with good wind resources could potentially supply up to **20 percent** of America's electricity.

Fossil Fuels Power Most Electricity Generation

As these graphs show, in 1973 and 2004 fossil fuels generated 75 percent and 66 percent, respectively, of the world's electricity. Oil consumption for electricity decreased from 1973 to 2004, while gas increased and coal remained about the same. Nuclear power generation increased significantly during the period, reducing dependency on fossil fuels.

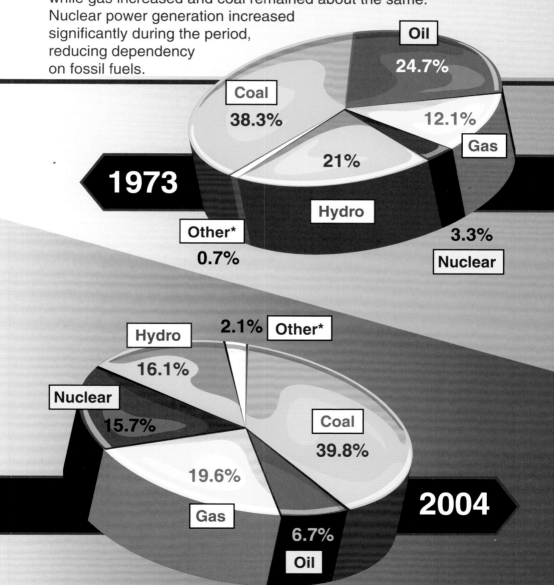

1973

Oil 24.7%

Coal 38.3%

12.1% Gas

21% Hydro

3.3% Nuclear

Other* 0.7%

2004

2.1% Other*

Hydro 16.1%

Nuclear 15.7%

19.6% Gas

Coal 39.8%

6.7% Oil

__Other__ includes geothermal, solar, wind, combustible renewables, and waste.

Source: International Energy Agency, "Key World Energy Statistics," 2006. www.iea.org.

Renewables Comprise a Small Percentage of America's Energy Supply

As this chart shows, in 2004 renewable energy comprised only 6 percent of the United States' energy supply. Hydroelectric and biomass energy made up most of this renewable energy.

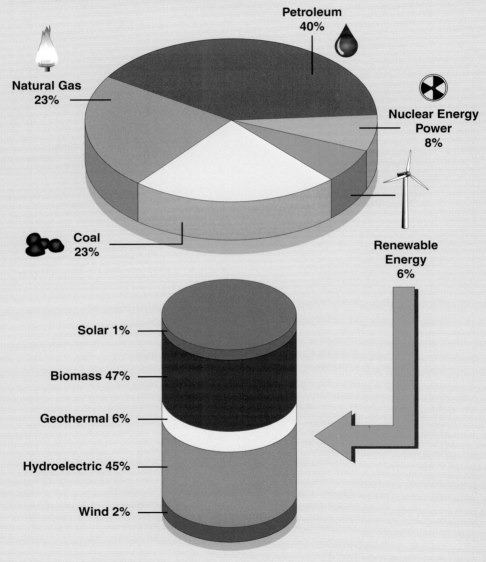

Petroleum 40%

Natural Gas 23%

Nuclear Energy Power 8%

Coal 23%

Renewable Energy 6%

Solar 1%

Biomass 47%

Geothermal 6%

Hydroelectric 45%

Wind 2%

Source: Energy Information Administration, "Renewable Energy Trends," August 2005. www.eia.doe.gov.

Africa Has the Greatest Solar Power Potential

This map shows the amount of solar energy, in hours, received each day on an optimally tilted surface during the worst month of the year. It reveals that the continents of Africa and Australia have the biggest potential for collecting solar energy.

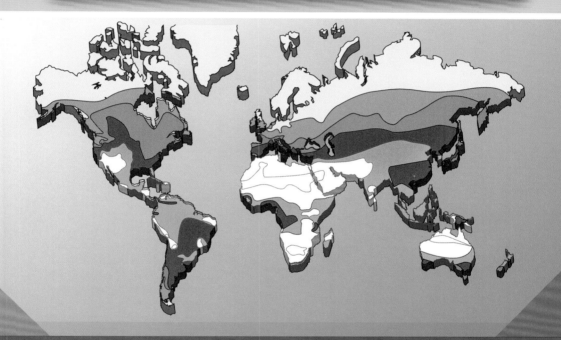

Watts/Square Meter

1.0–1.9		2.0–2.9		3.0–3.9	
4.0–4.9		5.0–5.9		6.0–6.9	

Wind Power Is Widespread in the United States

As this map of wind energy generation capacity shows, most states, except those in the southeastern United States, use wind power. Texas and California are the biggest wind energy producers, generating over 5,000 megawatts.

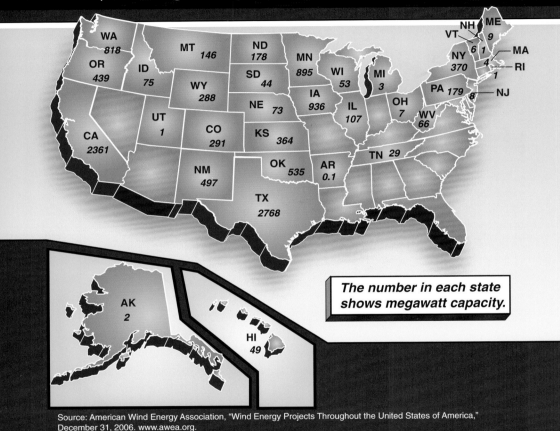

WA 818
OR 439
ID 75
MT 146
ND 178
MN 895
WI 53
MI 3
NH
VT
ME 9
NY 370
MA
RI
WY 288
SD 44
IA 936
PA 179
NJ 8
UT 1
CO 291
NE 73
IL 107
OH 7
WV 66
CA 2361
KS 364
TN 29
NM 497
OK 535
AR 0.1
TX 2768

The number in each state shows megawatt capacity.

AK 2
HI 49

Source: American Wind Energy Association, "Wind Energy Projects Throughout the United States of America," December 31, 2006. www.awea.org.

- According to the American Bird Conservancy, an average of **2.9 birds** are killed per turbine each year in wind farms in the United States.

- According to the Energy Information Administration, in 2004 hydropower dams accounted for about **45 percent** of total renewable energy consumption in the United States.

- The Energy Information Administration estimates that in 2004, geo-thermal energy accounted for about **6 percent** of total renewable energy consumption in the United States.

- According to the Energy Information Administration, because water is about **800 times** denser than air, tidal turbines will be much heavier and more expensive to build than wind turbines but will be able to capture more energy.

Can Alternative Energy Be Used for Transportation?

According to the Bureau of Transportation, in 2004 the total number of passenger vehicles in the United States was approximately 243 million. That year the U.S. population was estimated at close to 286 million. This amounts to almost one vehicle for each person living in the country. All these cars use a lot of fuel. According to the Central Intelligence Agency *World Factbook,* the United States consumes about 20.73 million barrels of oil per day but produces only 7.61 million barrels per day. In light of the fact that the United States and other nations demand so much more oil than they can produce, many people believe that society should look for alternative energies to power its transportation. Some of these potential energies are hydrogen power, hybrid-electric technology, and ethanol and other biofuels. Each alternative has positive and negative aspects, which cause heated disagreement concerning its potential.

The Potential of Hydrogen Power

Hydrogen is the most abundant chemical element in the world, and many people believe there is enormous potential to harness its power as a transportation fuel. According to the White House National Economic Council, "The Department of Energy estimated that, if hydrogen reaches its full potential . . . [it] could reduce our oil demand by over 11 million barrels per day by 2040—approximately the same amount of crude oil America imports today."[64] Hydrogen proponents point out that this fuel can be produced domestically, and that if it is produced using clean, renewable energy sources such as solar or wind power, it can reduce greenhouse gases and other pollution.

> " Hydrogen is the most abundant chemical element in the world, and many people believe there is enormous potential to harness its power. "

While hydrogen is not currently a viable fuel for widespread transportation use in the United States, many people believe technology is advancing rapidly enough to make it viable in the near future. In the opinion of journalist Drew Winter, "The hydrogen economy may not be as far away as most skeptics think, especially as fossil fuels grow more expensive and risky by the day."[65]

The United States is encouraging the development of hydrogen energy by allocating large amounts of money for hydrogen research. In 2003 president George W. Bush announced a $1.2 billion Hydrogen Fuel Initiative to develop commercially viable hydrogen power for transportation. Then in the 2007 budget, funding for hydrogen technologies was increased by $46 million over current levels. According to the White House National Economic Council's "Advanced Energy Alternative," "The promise of hydrogen technology is too great to ignore."[66]

Problems with Hydrogen for Transportation

However, hydrogen research faces numerous challenges. One is how to produce this fuel in a way that does not harm the environment. While hydrogen is the most abundant element in the universe, it takes energy to transform it into a usable form. Thus, hydrogen is only a clean source of energy if it is produced with clean sources of energy. Most hydrogen today is produced with oil and coal, and as journalist Joel Bainerman points out, "What good will generating energy from hydrogen do if the electricity for hydrogen generation comes from an existing dirty coal-fired power plant?"[67]

> **Transitioning to a hydrogen-based transportation system would involve changing the entire transportation infrastructure.**

In addition to the challenge of producing hydrogen, Bainerman points out that transitioning to a hydrogen-based transportation system would involve changing the entire transportation infrastructure and would thus be extremely expensive. He says, "That cost comprises establishing a completely new infrastructure to distribute hydrogen estimated to be at least $5,000 per vehicle, because transporting, storing and distributing a gaseous fuel as opposed to a liquid involves many difficult technical problems."[68]

Hybrid-Electric Technology

Hybrid-electric technology combines two energy sources—electricity and oil. A hybrid is similar to a conventional car except that its drive train has been modified to have both an internal combustion engine—which runs on gasoline, diesel, or an alternative fuel, and drives the front wheels—and one or two electric motors, powered by batteries, that drive the rear wheels. The electric motor shuts down during braking and coasting. Then the wheels drive the motor and the motor acts like a generator, charging the batteries.

There are differing opinions on how environmentally friendly hybrids are. Many people promote the technology because it can reduce oil consumption and because hybrids emit less air pollutants and greenhouse gases. However, critics caution that hybrids might not be as environmentally friendly as many people think. Journalist Chris Demorro argues that, contrary to popular belief, hybrids are a significant cause of pollution. Mining, refining, and transportation of the nickel needed to make the hybrid battery causes significant environmental damage, he says. For example, a nickel plant in Sudbury, Ontario, causes acid rain and sulfur dioxide pollution. "The area around the plant is devoid of any life for miles,"[69] says Demorro.

Owning a Hybrid-Electric Car

When considering whether to purchase a hybrid car, many people are influenced by the issue of price. For example, in 2007 Australia's *Cars Guide.com* asked, "Would you buy a hybrid-fuel car?" Reader Daniel responded, "Currently hybrid cars are too expensive to produce and buy. You have to clock up hundreds of thousands of k's [kilometers] to make it worthwhile. Factor in a few battery replacements (costing thousands of dollars each time) and it's completely pointless."[70] However the Web site *Hybrid Electric Cars* disagrees with such statements. It calls the belief that hybrids are too expensive a myth. Says the site, "While this may have been true just a few short years ago, it's no longer the case. Except for upscale models as expected. In early '06 the approximately ten models ranged from $18K to $55K. The most popular models—the Civic, Insight, and Prius—ran well under $30K."[71]

Another consideration in owning a hybrid is performance. In a 2007 article *CNNMoney.com* writer Peter Valdes-Dapena evaluates some common concerns of consumers thinking about purchasing hybrid cars. He explains that hybrid power trains are more efficient, so if you are comparing the same size engine, a hybrid might have better performance than a nonhybrid car. However, he says, if you are buying a hybrid in order to save on fuel, you should be choosing a smaller engine, so you probably will not get better performance than other cars. In summary, he says, the hybrid, "[Will] still offer power that's more than adequate for daily use, including merging and passing. You won't blow anyone's doors off, but you also won't waste quite as much of your life pumping gas."[72] As hybrid owner Christopher Todd points out, "Most people buy hybrids for fuel economy and to reduce pollution—not to blow the door off other cars." In Todd's opinion, hybrids offer excellent performance. He says, "I can say with certainty that this has been the most trouble-free vehicle I've ever owned."[73]

> " There are differing opinions on how environmentally friendly hybrids are. "

Biofuels

Biofuel is fuel derived from biomass—recently living organisms or their metabolic byproducts—such as plant matter or cow manure. The most common biofuel, ethanol, is produced from sugar or other starch crops. Biodiesel is less common and produced from oil-seed crops such as sunflowers. There is also current research on converting other types of biomass to create additional biofuels. According to the International Energy Agency (IEA), Brazil and the United States together account for approximately 80 percent of the world's ethanol supply. The United States manufactures ethanol primarily from corn, while in Brazil it is mainly produced from sugar cane. Most of the ethanol produced in the United States is blended in a low percentage with gasoline. According to Alexander Karsner, assistant secretary for the Office of Energy Efficiency and Renewable Energy, in 2005 ethanol provided approximately 3 percent of the United States' transportation fuel.

> **Most of the ethanol produced in the United States is blended in a low percentage with gasoline.**

There is disagreement over the potential of biofuels to meet transportation needs. In the opinion of Karsner, "Biofuels are the most promising near-term replacements for liquid transportation fuels."[74] The National Resources Defense Council maintains that using biofuels could actually save Americans about $20 billion per year by 2050. It says, "There's plenty of biomass around, and we can keep growing more of it."[75]

However, critics caution that biofuels are not a complete solution for transportation fuel. The IEA estimates that the production of biofuels currently uses about 1 percent of the world's arable land. "Given that 1% of global transportation fuels are currently derived from biomass, increasing that share to 100% is clearly impossible unless fuel demand is decreased . . . [or] land productivity is dramatically increased," it says, concluding, "For these reasons, the large-scale use of biofuels will probably not be possible unless . . . technologies . . . that [require] less arable land can be developed commercially."[76]

There is controversy over how increased use of biofuels could impact the environment. Critics worry about the effects of increasing overall agriculture and about intensive farming of a single crop. As conservationist Glen Barry explains, "Most terrestrial ecosystem destruction throughout history has resulted from agricultural conversion for farmland. Industrial agriculture causes water scarcity due to excessive irrigation, a toxic cocktail of agricultural chemicals which pervades food chains, and widespread soil erosion, infertility and even desertification."[77] Columnist George Monbiot says, "It is . . . an environmental disaster. Those who worry about the scale and intensity of today's agriculture should consider what farming will look like when it is run by the oil industry."[78] He believes that increased production of biofuels will destroy biodiversity. Biofuels expert Josh Tickell contends that the United States is not efficiently using its farmland and could easily grow more biofuels crops. "The United States has 60 million acres of fallow cropland we pay farmers not to grow on," he says. "There's massive potential to produce biofuels in this country."[79]

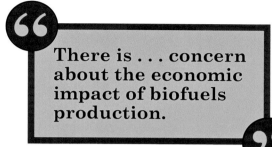

There is . . . concern about the economic impact of biofuels production.

There is also concern about the economic impact of biofuels production. According to editor Mike Clowes, increasing demand for corn for ethanol production has reduced America's corn exports, which has increased worldwide corn prices. Corn is used as feed for poultry, so such an increase has a far-reaching impact. For example, Clowes says, "The cost of raising chickens and eggs . . . has increased in many countries because of the higher cost of corn."[80]

However, others argue that biofuel technology will improve, allowing more efficient production of fuel and lessening economic impacts. For example, some ethanol researchers talk about cellulosic ethanol as a way to more efficiently produce large quantities of ethanol. Explains the Union of Concerned Scientists, "Cellulosic ethanol is more energy-efficient than corn ethanol and uses more abundant and diverse feedstocks that, unlike corn, are not used for food production."[81] However, cellulosic ethanol is not yet ready for commercial use.

Ethanol Use in Brazil

The nation of Brazil has made the greatest use of biofuels, and its actions have aroused both praise and criticism. Ethanol provides about 40 percent of automobile fuel there, and flex-fuel cars, which are able to run on either gasoline or ethanol, make up more than 70 percent of the new car market in Brazil. David Sandalow, director of the Environment & Energy Project at the Brookings Institution believes the impact of ethanol has been positive. He says, "The ethanol industry . . . takes credit for more than 1.8 million jobs in Brazil and for replacing, since 1976, more than 1.44 billion barrels of oil. . . . Ethanol contributes significantly to improving air quality in São Paolo and to cutting emissions of heat-trapping gases from the Brazilian transportation sector."[82]

However, Brazil's ethanol industry also has critics. According to André Kenji de Sousa, a resident of São Paulo, Brazil, the transformation to using ethanol has harmed the country in many ways. He says, "Large areas of land were wasted for monoculture, . . . semi-slave (and child) labor were heavily used. Nasty environmental problems . . . like the pollution of rivers by vinhoto (produced in ethanol refining). . . . Ethanol apologists shouldn't be talking about 'Brazilian ethanol success' because there is no such success."[83]

> **Ethanol provides about 40 percent of automobile fuel [in Brazil].**

The different alternative transportation fuels obviously have numerous problems as well as benefits. In light of such pros and cons, there is disagreement over whether they should be developed to sustain transportation systems. Tickell says, "[The alternative to oil is] not perfect; it never will be. But . . . until people get realistic about that choice, they'll continue to kid themselves that something better is around the corner. Something better has arrived; it's here."[84] Researchers Jerry Taylor and Peter Van Doren contend that until transportation energy alternatives decrease in price and increase in efficiency, oil will continue to dominate as a transportation fuel because consumers will choose the lowest-cost source of energy. However, as the price of oil rises and concerns about fossil fuels increase, society is increasingly looking to energy alternatives in transportation and debating whether these alternatives—hydrogen, hybrid-electric technology, and biofuels—can replace oil.

Primary Source Quotes*

Can Alternative Energy Be Used for Transportation?

66Non-petroleum fuels are becoming viable and necessary alternatives to gasoline and diesel fuels.**99**

—California Energy Commission, "Integrated Energy Policy Report," November 2005. www.energy.ca.gov.

The California Energy Commission is the state's primary energy policy and planning agency. It keeps historical energy data, forecasts future energy needs, and promotes energy efficiency and the development of new energy technologies.

66Unfortunately, there is nothing on the horizon that comes close to gasoline as far as cost and performance is concerned. . . . For the time being . . . cheap fuel means gasoline.**99**

—Jerry Taylor, "For Now, Gasoline Is Our Only Cheap Fuel," *Arizona Republic,* May 7, 2006. www.azcentral.com.

Taylor is a senior fellow at the Cato Institute where he researches environmental policy.

Bracketed quotes indicate conflicting positions.

* Editor's Note: While the definition of a primary source can be narrowly or broadly defined, for the purposes of Compact Research, a primary source consists of: 1) results of original research presented by an organization or researcher; 2) eyewitness accounts of events, personal experience, or work experience; 3) first-person editorials offering pundits' opinions; 4) government officials presenting political plans and/or policies; 5) representatives of organizations presenting testimony or policy.

> **66Auto companies would be well advised to put in high gear their research into smaller, lighter, and more energy-efficient electric, hybrid, and even human-powered vehicles.99**

—Richard Heinberg, *The Party's Over: Oil, War, and the Fate of Industrial Societies.* Gabriola Island, BC: New Society, 2003, p. 232.

Heinberg is an award-winning author and an expert on energy resources.

> **66[New] technologies may ultimately shift our primary transportation fuel from petroleum . . . to hydrogen.99**

—David Garman, testimony before the Committee on Energy and Natural Resources, July 17, 2006.

Garman is the U.S. undersecretary of energy.

> **66To get serious about energy policy, America needs to abandon, once and for all, the false promise of the hydrogen age.99**

—Robert Zubrin, "The Hydrogen Hoax," *New Atlantis,* Winter 2007. www.thenewatlantis.com.

Zubrin, an aerospace engineer, is president of Pioneer Astronautics, a research and development firm. His new book on energy policy will be published in the fall of 2007.

> **66It should be national policy to foster early introduction on a significant scale of [alternative] vehicle technologies and non-petroleum transportation fuels for they promise to make a major contribution to U.S. energy security.99**

—R. James Woolsey, testimony before the U.S. House of Representatives Committee on Government Reform, Subcommittee on Energy and Resources, April 6, 2005. www.energycommission.org.

Woolsey is a member of the National Commission on Energy Policy and a former director of the Central Intelligence Agency.

"My Administration is committed to conserving America's public lands and natural resources . . . and encourage[s] the use of hybrid cars. . . . Through efforts like these, we will . . . protect the environment."

—George W. Bush, "Proclamation 8023—Great Outdoors Month, 2006," May 23, 2006. www.whitehouse.gov.

Bush is the forty-third president of the United States.

"Unfortunately . . . [the] ultimate 'green' car is the source of some of the worst pollution in North America; it takes more combined energy per Prius [hybrid] to produce than a Hummer."

—Chris Demorro, "Prius Outdoes Hummer in Environmental Damage," *Recorder*, March 7, 2007. http://clubs.ccsu.edu.

Demorro is a staff writer for the Central Connecticut State University *Recorder*.

"Biofuels hold out the prospect of replacing substantial volumes of imported oil . . . [and] could also bring environmental benefits."

—International Energy Agency, *World Energy Outlook 2006*. Paris, France: International Energy Agency, 2006, p. 386.

The International Energy Agency is an intergovernmental organization dedicated to preventing disruptions in the supply of oil as well as acting as an information source on statistics about the international oil market and other energy sectors.

"There is no way enough biomass could ever be grown to meet a significant portion of current, much less anticipated, world energy needs without causing great environmental harm."

—Glen Barry, "Bursting Biofuels' Bubble," *Earth Meanders*, April 25, 2006. http://earthmeanders.blogspot.com.

Barry, a conservation biologist and political ecologist, writes essays and Internet blogs.

66My Civic Hybrid has been the best car I've ever owned. . . . Overall, I'd give my car a satisfaction rating of 9 out of 10.99

—Christopher Todd, "My Civic Hybrid Experience!" March 18, 2007. www.gaianar.com.

Todd is a hybrid owner who lives in Baltimore, Maryland.

66Reports that indicate that corn ethanol production provides a positive [energy] return . . . mislead . . . U.S. policy makers and the public. . . . [Our research shows] a negative 29% deficit.99

—David Pimentel and Tad W. Patzek, "Ethanol Production Using Corn, Switchgrass, and Wood; Biodiesel Production Using Soybean and Sunflower," *Natural Resources Research,* March 2005, pp. 65, 69.

Pimentel is a researcher at Cornell University, New York, and Patzek is a researcher at the University of California at Berkeley.

66Critics claim the production of ethanol is not efficient. . . . But nine major studies done since 1995 show that corn ethanol contains, on average, about 30 percent more energy than the fossil fuels needed to make it.99

—Samuel Bodman, "Ask the White House," January 24, 2007. www.whitehouse.gov.

Bodman is the U.S. secretary of energy.

❝I have a farm. I like the fact that my farm neighbors can put their harvested corn on a truck . . . and have it turned into fuel. The fuel isn't shipped from half way around the world, and my neighbors are seeing the best corn prices they have seen in 10 years.❞

—John Bondi, "Environmental Claims Should Be Based on the Real World, Not Models," *Wisconsin Technology Network,* August 4, 2006. http://wistechnology.com.

Bondi is a columnist for the *Wisconsin Technology Network,* a news organization that works to connect the people, technology, and ideas driving the advancement of life sciences, biotechnology, and information technology in Wisconsin.

Can Alternative Energy Be Used for Transportation?

- According to the Sierra Club, cars and light trucks emit **20 percent** of America's carbon dioxide pollution.

- According to a 2005 report by the Energy Information Administration, **92 percent** of American households own or possess one or more vehicles.

- In 2006 the *San Francisco Chronicle* reported that a hydrogen fuel cell car cost **$1 million**.

- According to the U.S. Department of Energy, when compared to conventional vehicles using gasoline, fuel cell vehicles using hydrogen produced from natural gas reduce greenhouse gas emissions by **60 percent**.

- *Hybrid Car Review* reports that the number of hybrid vehicles sold in the United States increased **245 percent** between 2004 and 2005, from 83,924 to 205,748.

- According to a 2007 *Los Angeles Times* report, if hybrid cars replaced all **245 million** cars in the United States, the carbon dioxide savings would be less than 3 percent of what is needed to reduce global warming.

- The National Resources Defense Council finds that biomass energy now accounts for **45 percent** of the renewable energy used in the United States.

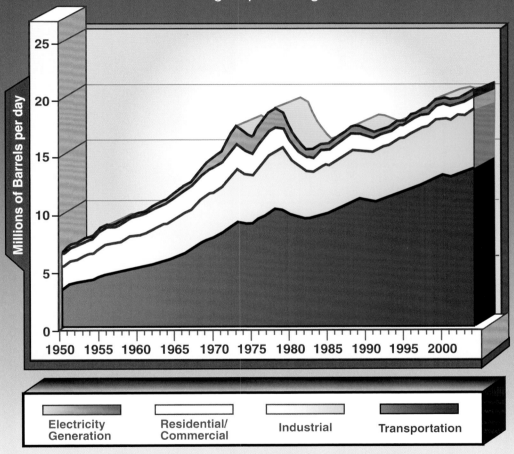

Majority of U.S. Petroleum Consumption Is for Transportation

This chart shows that most of America's oil—more than 60 percent—is used for transportation. Industry uses the next largest percentage of oil.

Millions of Barrels per day

1950 1955 1960 1965 1970 1975 1980 1985 1990 1995 2000

Electricity Generation | Residential/ Commercial | Industrial | Transportation

Source: Energy Information Administration, "U.S. Oil Demand by End-Use Sector, 1950–2004." www.eia.doe.gov.

- According to the White House National Economic Council, in 2004 ethanol comprised **2 percent** of the volume of all gasoline sold in the United States.

- Brazil meets approximately **20 percent** of its liquid transportation needs with ethanol.

- According to Alexander Karsner, assistant secretary for the Office of Energy Efficiency and Renewable Energy, the most ethanol that could be produced from corn on a sustainable basis in the United States is **13 percent** of transportation fuel.

U.S. Hybrid Sales Are Increasing

According to this graph of hybrid car sales for 2004–2007, each year, sales are higher than the year before, with the largest increase occurring in the beginning of 2007.

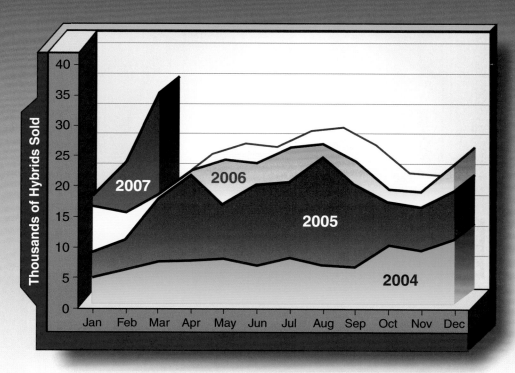

Source: Green Car Congress, "Reported Sales of Hybrids Almost Double in March 2007; 2.25% Share of New Vehicle Sales," April 6, 2007. www.greencarcongress.com.

Brazil and the United States Lead the World in Ethanol Production

This chart shows that Brazil and the United States are the world's primary ethanol producers. While in 2000 Brazil produced significantly more ethanol than the United States, in 2005 the two countries produced about the same. The European Union, China, and India also produce small amounts of ethanol.

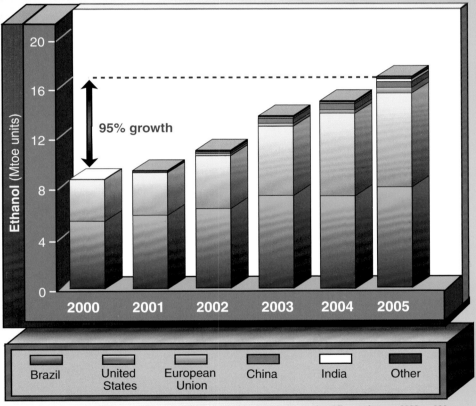

Source: International Energy Agency, *World Energy Outlook 2006*. France: International Energy Agency, 2006, p. 390.

- According to the National Corn Growers Association, ethanol production results in a net energy gain, producing **67 percent** more energy than it takes to grow and process the corn into ethanol.

- The U.S. Department of Agriculture estimates that nearly **20 percent** of the country's 2006 corn crop will be converted into ethanol.

Safety, Cost, and Convenience Are Consumers' Biggest Transportation Concerns

According to this poll of 1,006 American adults, the most important consumer concerns in relation to fuel technology are safety, the cost of fuel and the vehicle, and the convenience of refueling.

Factors Considered "Extremely Important" or "Very Important" in Influencing Decisions to Try a New Fuel Technology

Factor	Percent
How safe the fuel is for drivers and passengers	83%
The cost of the fuel	78%
How far you can drive before refueling	75%
The cost of the vehicle	72%
The convenience of refueling	67%
Environmental emissions	67%
Whether the fuel source is domestic instead of foreign	47%
How the new fuel system affects passenger and cargo space	47%
Whether or not the fuel can be recycled	45%

Source: M. Kubrik, "Consumer Views on Transportation and Energy (Third Edition)," *National Renewable Energy Laboratory*, January 2006. www.nrel.gov.

How Will Increased Use of Energy Alternatives Impact Society?

> 66 Renewable sources like wind, solar, and geothermal energy . . . would do more than just cut pollution and curb global warming. . . . [They] would jump start our economy. 99

—Sierra Club, "Global Warming: A Time for Action."

> 66 Any . . . transition [to energy alternatives] will be costly— in terms of dollars, energy, and/or our standard of living. . . . It is misleading to think we can achieve . . . [it] easily or painlessly. 99

—Richard Heinberg, *The Party's Over: Oil, War, and the Fate of Industrial Societies.*

Many people believe that due to problems with the current energy supply, it is time for society to dramatically increase its use of alternatives. However, there is uncertainty about how an expanded role for alternatives will impact society. While some people focus on the advantages of alternatives, others point out the problems.

Writer Ben Hewitt cautions, "It's important that we don't turn a blind eye to potential downsides."[85] In the opinion of Hewitt and others, increased use of alternatives will cause dramatic changes in the way society uses energy. For example, Hewitt believes that increased conservation will be necessary. As the use of energy alternatives begins to increase, the debate about how society will be affected continues. Topics of discussion include how reduced oil use will impact society and how increased use of energy alternatives will impact the environment, national security, the economy, and overall energy use patterns.

Overall Impact of Reduced Use of Oil

Most scenarios involving increased use of energy alternatives assume that society will also begin to use less fossil fuel. Many people take an optimistic view of such a transition. For example, energy researcher Barry J. Hanson, who lives in an alternative energy home and works in an alternative energy business facility, insists that a power shift would be extremely beneficial. He maintains, "The transition [to a renewable energy economy] . . . will have far reaching political, social and economic implications—creating millions of jobs, revitalizing local economies, reducing the federal trade deficit."[86]

Yet others, such as energy expert Richard Heinberg, give a pessimistic prediction of any transition from fossil fuels to alternative fuels. In Heinberg's opinion, fossil fuels will run out, so there must be a transition to alternatives, but he believes that with such a transition it will be impossible for society to continue as it is. He points out that society has been built on the use of increasing amounts of fossil fuels, and the current way of life depends upon a continuing supply of those fuels. He believes that without fossil fuels, "Global societal collapse . . . is likely, and perhaps certain."[87] In Heinberg's opinion, decreasing use of fossil fuels will seriously impact transportation, the food supply and agricultural system, public health, and the economy. He says that without fossil fuels, there will necessarily be fundamental changes in the way people live; for example, there will be less travel, and people will need to live in smaller communities, use less energy, and produce food locally.

> **Society has been built on the use of increasing amounts of fossil fuels.**

Environmental Benefits

Proponents of alternative energy point out that fossil fuels often cause significant harm to the environment and human health. They believe alternatives can benefit society by substantially reducing these harms. According to the United Nations and the World Energy Council, "The environmental impacts of a host of energy-linked emissions . . . contribute to local and regional air pollution and ecosystem degradation. Human health is threatened by high levels of air pollution."[88]

In contrast, many alternative forms of energy generation such as solar, wind, and geothermal technologies do not release harmful pollutants into the environment. While advocates admit that energy alternatives also have an environmental impact, they believe it is far more desirable than that of fossil fuels. As *Audubon* contributor John Flicker explains, "Every source of energy has some environmental consequences. . . . Our challenge is thus to help design and locate . . . projects that minimize the negative impacts."[89]

> " Many alternative forms of energy generation . . . do not release harmful pollutants into the environment. "

Potential Environmental Harms

Skeptics of energy alternatives insist that these various energy technologies can also cause significant environmental harm. In the opinion of the National Center for Policy Analysis, "Renewable energy sources are not as environmentally benign as their promoters claim. Each source of renewable energy produces negative environmental side-effects."[90] For example, says the center, water from geothermal reservoirs often contain minerals that are corrosive and polluting, and wind farms produce noise pollution and kill birds.

According to Bill Peacock, director of the Center for Economic Freedom at the Texas Public Policy Foundation, the environmental benefits of renewable energy are often overstated. He uses the example of wind power to make his point:

Wind power often cannot produce electricity when needed

because the wind isn't blowing. This means fossil fuel power stations must continually run as a backup. Combining this with the pollutants emitted in the manufacture and maintenance of wind towers and their associated infrastructure means that substituting wind power for fossil fuels provides limited improvements in air quality.[91]

There is also skepticism concerning the environmental impact of biofuels. Among other things, critics worry that intensive farming of biofuel crops will reduce biodiversity, contribute to deforestation, and also cause environmental harm through increased use of pesticides.

National Security

Many people believe that dependence on fossil fuels is a big threat to national security and that increasing America's use of energy alternatives can reduce oil use and enhance national security. For example, energy expert Howard Geller discusses the link between oil and terrorism. He says, "The terrorist attacks on the World Trade Center and Pentagon were not unconnected to our high oil import dependence. Revenue from oil sales finance terrorist groups such as the al-Qaeda network. . . . [Terrorist] Osama bin Laden was enraged by the presence of American soldiers in Saudi Arabia, a presence driven by our thirst for oil."[92] Illinois senator Barack Obama urges America to reduce its oil use by increasing investment in alternative, domestically produced forms of energy. He says, "For all of our military might and economic dominance, the Achilles heel of the most powerful country on Earth is the oil we cannot live without. . . . [We need to] harness our own renewable forms of energy so that oil can never be used as a weapon against America."[93]

> " Skeptics of energy alternatives insist that these various energy technologies can also cause significant environmental harm. "

Yet critics caution that it is impossible to completely eliminate energy imports. As the United States Energy Association points out, "U.S. energy security cannot be achieved by closing our border to energy imports or by limiting energy exports. Attempts to do so would cripple the economy, limit trade, slow the creation of wealth around the globe, and delay the spread of technology (and, oftentimes, open markets) to developing nations."[94] According to this reasoning, the United States will always be dependent on imported energy to some degree and thus vulnerable to the actions of other nations.

Economic Impact

Many advocates of energy alternatives insist that increased use of these technologies will increase employment and greatly benefit the economy. In the opinion of J. Andrew Hoerner of the Center for a Sustainable Economy and James Barrett of the Economic Policy Institute, "For too long, the debate over America's energy future has been hamstrung by the outdated notion that there's an inherent tradeoff . . . for example, that we can reduce oil imports and stop global warming—or we can have job growth."[95] This is a false belief, insist the authors. They argue that if America invests in renewable energy, the economy will actually

> **Critics caution that it is impossible to completely eliminate energy imports.**

be revitalized and many new jobs created. In addition, they maintain that American consumers will end up saving money. For example, in June 2005 the U.S. Senate passed an energy bill requiring all large electric utilities to gradually increase their use of renewable energy sources to at least 10 percent by 2020. A study by the Energy Information Administration found that this would foster creation of over 190,000 jobs, providing the U.S. economy with an additional $5.1 billion in income and $5.9 billion in gross domestic product.

Some people insist that energy alternatives will not only benefit the economy, but that it will actually be economically harmful if the United States does not invest in the alternative energy market. For example, according to a 2005 report by the National Resources Defense Council, U.S. carmakers are failing to take advantage of demand for gas-electric hybrid cars. The organization believes that as a result, the United States will lose thousands of jobs to foreign automakers who are producing these cars.

Critics like Michael Kane, an associate administrator at the U.S. Department of Energy, suggest that the opposite will be true; that the economy will deteriorate rather than be revitalized by alternative energies. He says, "Renewable projects should be developed, but we must be realistic about what they can and cannot provide to society."[96] Kane and others believe that if society replaces oil with alternative forms of energy, the economy will be forced to slow down. He argues that there is only one solution, and that is "the intentional decrease of energy consumption by the entire industrial world."[97]

Local Economies

In relation to the economic impact of energy alternatives, many advocates for alternatives believe that local economies in particular will benefit. The American Solar Energy Society explains, "Every dollar spent on energy imports is a dollar that the local economy loses. Renewable energy resources, however, are developed locally. The dollars spent on energy stay at home, creating more jobs and fostering economic growth."[98] Undersecretary for Rural Development Thomas C. Dorr argues that the development of renewable energy represents an historic opportunity for job and wealth creation in rural America. He explains that small and mid-sized producers of biodiesel, wind, and solar power can compete in such a market, whereas such competition is extremely difficult in the fossil fuels market.

> **Many advocates of energy alternatives insist that . . . these technologies will increase employment and greatly benefit the economy.**

With oil as the major energy source, many nations such as the United States have become increasingly dependent on the few regions that have it, such as the Middle East. A shift to locally produced sources such as wind and solar power would be a reversal of this trend.

The Cost of Using Alternatives

Extensive debate continues over whether alternative energies are actually affordable for society to use. Some people insist that they are too expensive. Says journalist Jim McKay, "The simple fact is, alternative energy in all its forms is not yet competitive on the price front for everyday users. It only works, even advocates for environmentally friendly energy say, with the help of government subsidies or good corporate citizens who are willing to pay a premium so they can market themselves as being 'green.'"[99]

Others contend that alternative energy is actually cheaper than most people realize. The Union of Concerned Scientists explains:

> **Some people insist that [energy alternatives] are too expensive [to develop].**

> Some energy costs are not included in consumer utility or gas bills, nor are they paid for by the companies that produce or sell the energy. These include human health problems caused by air pollution from the burning of coal and oil; damage to land from coal mining and to miners from black lung disease; environmental degradation caused by global warming, acid rain, and water pollution; and national security costs, such as protecting foreign sources of oil.[100]

According to the organization, "Since the producers and the users of energy do not pay for these costs, society as a whole must pay for them. But this pricing system masks the true costs of fossil fuels and results in damage to human health, the environment, and the economy."[101]

Energy Conservation

Many people believe that because energy alternatives generally provide less energy than oil, greater reliance on these alternatives will necessarily lead to energy conservation. In the opinion of Montana resident William von Brethorst, "Here is the problem with renewable energy. Our present day energy use is so large (about 907 gigawatts in the USA each day) that no amount of renewable energy will ever be able to replace the conventional power sources without energy conservation on a massive scale."[102] Thus Brethorst and others believe that society will be forced to use less energy overall. According to the Schatz Energy Research Center, such conservation will be a critical part of the success of alternatives. It argues,

"The more energy-efficient we make our society, the more feasible it will be to convert to an all-renewables economy."[103]

Researcher Jim Holm contends that society must find an energy solution that provides as much energy as from oil. Current alternatives are simply not an adequate solution, he insists. In his opinion, society depends heavily on reliable, high-energy sources to power necessities such as utilities and transportation. He says, "This is why, when it comes to the electricity that powers our mega-cities, clean mega-solutions are the solutions that deserve top priority. . . . Think about Baghdad and the chaos its intermittent electricity creates."[104]

> Many people believe that . . . greater reliance on . . . alternatives will necessarily lead to energy conservation.

In the opinion of Audubon Society writer Cheryl Coon, "It would be wise to remember that every source of energy has benefits and costs. Facing that reality honestly . . . will be critical to our ability to achieve a healthy planet for ourselves and for future generations of all its inhabitants."[105] As Coon's advice highlights, whether positive or negative, greater use of energy alternatives will definitely impact society in many ways. However, no one can be sure exactly how, and policy makers, researchers, and the public continue to disagree on the potential effects of reduced oil use and the impacts of alternative energy use on the environment, national security, the economy, and overall energy consumption.

How Will Increased Use of Energy Alternatives Impact Society?

66By . . . growing reliance on renewable energy sources . . . all of the problems associated with current energy patterns and trends can be mitigated.99

—Howard Geller, *Energy Revolution: Policies for a Sustainable Future.* Washington, DC: Island, 2003, p. 2.

Geller is executive director of the Southwest Energy Efficiency Project, a program that promotes policies and programs to advance energy efficiency in Arizona, Colorado, Nevada, New Mexico, Utah, and Wyoming.

66After [oil runs out] . . . nation-states break down, the frantic attempts of people to feed themselves, stay warm and obtain fresh water . . . there will be no rescue. Die-off begins. . . . What about renewable energy and other alternatives? They are not ready. . . . When oil abdicates, no one can fill the shoes.99

—Jan Lundberg, "Here Comes the Nutcracker: Peak Oil in a Nutshell," *Energy Bulletin*, February 18, 2005. www.energybulletin.net.

Lundberg is an environmental activist and an oil industry analyst.

WITHDRAWN

Bracketed quotes indicate conflicting positions.

* Editor's Note: While the definition of a primary source can be narrowly or broadly defined, for the purposes of Compact Research, a primary source consists of: 1) results of original research presented by an organization or researcher; 2) eyewitness accounts of events, personal experience, or work experience; 3) first-person editorials offering pundits' opinions; 4) government officials presenting political plans and/or policies; 5) representatives of organizations presenting testimony or policy.

❝Probably the most significant social benefit that a renewable approach would bring is relative to the environment. From global warming to acid rain and nuclear waste, a whole host of environmental nightmares would be eliminated.❞

—Barry J. Hanson, *Energy Power Shift: Benefiting from Today's New Technologies.* Maple, WI: Lakota Scientific, 2004, p. 10.

Hanson is an energy expert who lives in an alternative energy home and works in an alternative energy business facility, both of which he designed.

❝The clear and present danger to our . . . security from America's long-term dependency on oil will not subside—unless we act now . . . by using the kind of clean, renewable sources of energy that we can literally grow right here in America.❞

—Barack Obama, speech to Resources for the Future, September 15, 2005. http://obama.senate.gov.

Obama is a U.S. senator representing Illinois.

❝Energy independence is simply unrealistic. . . . U.S. oil imports will continue to grow in the future, as they have for the last several decades, and . . . we like everyone else will increasingly need to rely on oil supplies that originate in what are now unstable and undemocratic regions of the world.❞

—Jason Grumet, congressional testimony, Hearing on Energy Security and Oil Dependence, Senate Committee on Foreign Relations, May 16, 2006. http://energycommission.org.

Grumet is executive director of the National Commission on Energy Policy.

❝With 16 ethanol and biodiesel plants, Minnesota has a lot to boast about. All of Minnesota's 2.6 billion gallons of gasoline are blended with 10 percent ethanol, creating 40,000 jobs for Minnesotans and contributing $500 million in economic activity each year.❞

—Norm Coleman, "Expanding the Use of Renewable Fuels," March 16, 2005. http://coleman.senate.gov.

Coleman is a U.S. senator representing Minnesota.

❝In order to finance an aggressive implementation of alternative energies, we need a tremendous amount of investment capital. . . . We [also] need 25-to-50 years to retrofit our economy to run on alternative sources of energy.❞

—Matthew David Savinar, "Won't High Oil Prices Motivate Us to Look for Alternatives?" *Life After the Oil Crash*, January 2007. www.lifeaftertheoilcrash.net.

Savinar is an attorney and oil expert who has appeared on numerous national and international radio shows to discuss the implications of a declining oil supply.

❝The economic cost of . . . [alternative energy] policies would be more than outweighed by the economic benefits that would come from using and producing energy—more efficiently.❞

International Energy Agency, *World Energy Outlook 2006*. Paris, France: International Energy Agency, 2006, p. 37.

The International Energy Agency is an intergovernmental organization dedicated to preventing disruptions in the supply of oil as well as acting as an information source on statistics about the international oil market and other energy sectors.

❝[Our analysis of Colorado found that] investments in renewable energy, dollar for dollar, produce a greater net benefit for Colorado's economy than traditional technologies.❞

—Travis Madsen, Timothy Telleen-Lawton, Will Coyne, and Matt Baker, "Energy for Colorado's Economy: Creating Jobs and Economic Growth with Renewable Energy," *Environment Colorado,* February 2007. www.environmentcolorado.org.

Madsen and Telleen-Lawton are policy analysts for the Federation of State Public Interest Research Groups. Coyne is a land use advocate, and Baker is executive director for Environment Colorado.

❝Alternatives to old and dirty fossil fuels are now business-friendly . . . [and] cost-competitive.❞

—National Resources Defense Council, "Wind, Solar, and Biomass Energy Today," January 12, 2006. www.nrdc.org.

The National Resources Defense Council is a nonprofit organization of lawyers, scientists, and environmentalists dedicated to protecting the environment and public health.

❝Alternative [energy] sources . . . will require additional advancement before they can become truly cost-competitive with conventional energy sources.❞

—Environmental Literacy Council, "Energy," November 14, 2006. www.enviroliteracy.org.

The Environmental Literacy Council is a nonprofit organization that gives teachers the tools to help students develop environmental literacy.

❝To go from our fossil-fuel world to a biomass world would be a little like leaving the Garden of Eden for the land where bread must be earned by 'the sweat of your brow.'❞

—Bill McKibben, "Reversal of Fortune," *Mother Jones,* March/April 2007. www.motherjones.com.

McKibben is an environmentalist who has written numerous articles on global warming and alternative energy.

Facts and Illustrations

How Will Increased Use of Energy Alternatives Impact Society?

- Argonne National Laboratory finds that the use of **10 percent** ethanol fuel blends reduces greenhouse gas emissions by **18 to 29 percent** compared with conventional gasoline.

- Analysis by the National Center for Policy Analysis shows that even if every source of human-caused carbon dioxide emissions were eliminated overnight, the atmosphere would continue to warm for **100 years** or more.

- According to the National Defense Council Foundation, America's oil dependence results in **$297 to $305 billion** each year in costs such as loss of jobs and tax revenues.

- According to the American Solar Energy Society, the United States manufactures about two-thirds of the world's photovoltaic systems—systems that convert sunlight to energy—resulting in annual exports worth more than **$300 million**.

- The Union of Concerned Scientists estimates that increasing renewable electricity use to **20 percent** by 2020 would save consumers nearly $27 billion.

- The National Corn Growers Association finds that a typical **40-million-gallon** ethanol plant creates **32 full-time jobs** and generates an additional **$1.2 million** in tax revenue for the community.

Renewable Energy Costs Are Declining

These graphs show that since 1980 the costs of various renewable energies have all declined dramatically. The graphs indicate that this trend is expected to continue in the future.

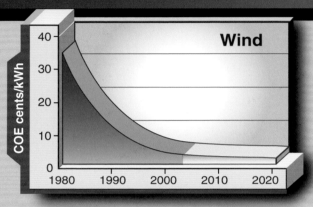

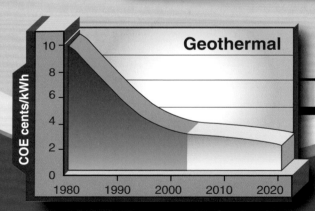

COE = Cost of Energy

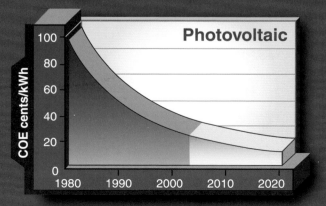

How Will Increased Use of Energy Alternatives Impact Society?

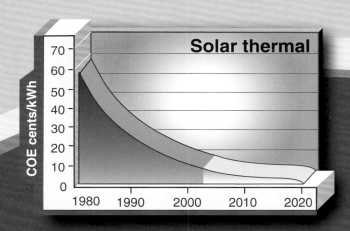

These graphs are reflections of historical cost trends, NOT precise annual historical data.

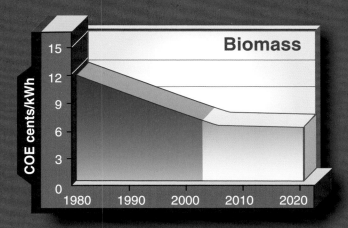

Source: Michael Eckhart, "Renewables: Looking Toward Inexhaustible Energy," *eJournal USA: Economic Perspectives*, July 2006. http://usinfo.state.gov.

Widespread Public Concern for the Social Impact of Continuing Fossil Fuel Use

This 2006 poll of 19,579 people in 19 different countries found that there is widespread public concern that the current fossil-fuel dominated energy supply will have a harmful effect on the environment, the economy, and international relations.

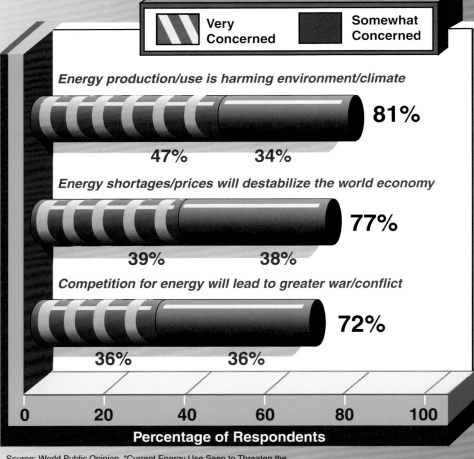

Very Concerned   **Somewhat Concerned**

Energy production/use is harming environment/climate
81%
47%   **34%**

Energy shortages/prices will destabilize the world economy
77%
39%   **38%**

Competition for energy will lead to greater war/conflict
72%
36%   **36%**

0   20   40   60   80   100

Percentage of Respondents

Source: World Public Opinion, "Current Energy Use Seen to Threaten the Environment, Economy, Peace," July 12, 2006. www.worldpublicopinion.org.

- According to the Renewable Fuels Association, in 2005 the ethanol industry supported the creation of more than **153,725 jobs** in all sectors of the U.S. economy, boosting U.S. household income by **$5.7 billion**.

- Analysis by the American Wind Energy Association shows that average small wind turbine costs have **decreased** by 7 percent since 2002 and that manufacturers are aiming to reduce costs another **20 percent** by 2010.

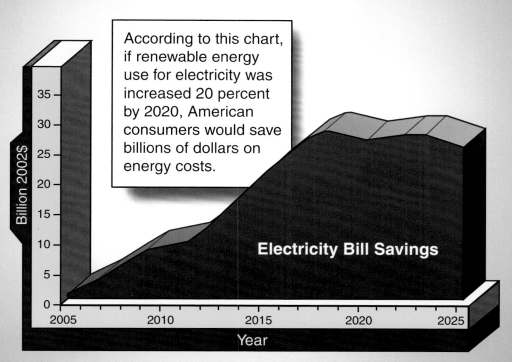

Renewable Electricity Would Save Consumers Money

According to this chart, if renewable energy use for electricity was increased 20 percent by 2020, American consumers would save billions of dollars on energy costs.

Electricity Bill Savings

Source: Union of Concerned Scientists, "Renewable Energy Can Help Ease Natural Gas Crunch," August 26, 2005. www.ucsusa.org.

Government Funds for Renewable Energy Have Increased

This graph shows the U.S. government's energy-related budget requests between 1998 and 2007. It reveals that in recent years, fossil fuel and energy conservation program funding has decreased, while funds for renewable energy have increased.

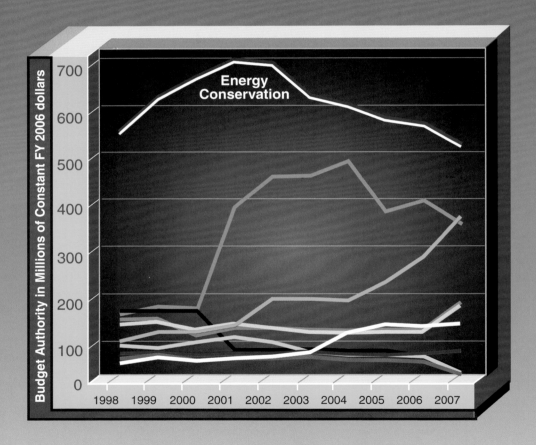

Non-renewable

- Coal
- Oil
- Nuclear
- Gas

Renewable

- Energy Biomass
- Solar
- Wind
- Geothermal
- Hydrogen

Source: American Association for the Advancement of Science, "DOE Science Gains 14 Percent, Energy R&D Slides in 2007 Budget," February 21, 2006. www.aaas.org.

- According to the Renewable Energy Policy Network for the 21st Century, in 2005 approximately **$38 billion** was spent on renewable energy development around the world.

- The White House National Economic Council reports that the 2007 presidential budget includes **$281 million** for coal research, **$148 million** for solar energy, and **$44 million** for wind energy.

Key People and Advocacy Groups

Samuel Bodman: Bodman is the U.S. secretary of energy and leads the Department of Energy.

David Bransby: Bransby is a nationally recognized authority on biofuels, and part of the advisory board for Earth Biofuels, Inc.

Steve Clemmer: Steve Clemmer is senior energy analyst for the Union of Concerned Scientists, a nonprofit organization that combines scientific research and citizen action to promote a healthy environment.

Bradley D. Collins: Collins is executive director of the American Solar Energy Society. He works to educate and advocate for renewable energy technologies.

Karl Galwell: Galwell is executive director of the Geothermal Energy Association. Before that he was the director of government affairs for the American Wind Energy Association.

David Garman: David Garman is the U.S. undersecretary of energy.

Jeremy Leggett: Leggett is an award-winning scientist, oil industry consultant, and Greenpeace campaigner.

Ken Sheinkopf: Ken Sheinkopf is director of public affairs at the Florida Solar Energy Center. Since 1986 he has written a weekly column on home energy that is nationally syndicated to more than 400 newspapers and magazines.

James Sims: Sims is executive director of the American Green Network, a sustainable energy advocacy organization. He has worked as a renewable energy advocate and lobbyist since 1992.

Jaime Steve: Steve is legislative director of the American Wind Energy Association, a national trade association that promotes wind energy for electricity.

Josh Tickell: Tickell is a pioneer in biofuels development. He has consulted for companies such as the National Biodiesel Board and has written two books about biodiesel.

Chronology

1950
Over a third of all U.S. houses are warmed by coal. Natural gas is used to warm about 25 percent of U.S. households.

pre-1885
Wood is the primary energy source for cooking, warmth, light, trains, and steamboats.

1954
Researchers develop the first solar cell capable of generating enough power from the sun to run everyday electrical equipment.

1945–1978
After World War II ends, with reduced need for war materials and with the low price of fuel, ethanol use as a fuel is drastically reduced.

1888
The first large windmill is used to generate electricity in Cleveland, Ohio. The windmill starts to be called the "wind turbine."

1885 1890   1910   1930   1950   1970

1885–1950
Coal is the primary source of fuel.

1957
The first commercial nuclear power plant begins operating.

1890
Mass production of automobiles begins, creating demand for gasoline. Prior to this, kerosene used for lighting had been the main oil product.

1970
U.S. production of petroleum reaches its highest level at 11.7 million barrels per day. Overall production has generally declined since then.

1917–1918
The need for fuel during World War I drives up ethanol demand significantly.

1970s
A significantly less-costly solar cell is developed, bringing the price down from $100 per watt to $20 per watt.

1973
An oil embargo by the Organization of Petroleum Exporting Countries (OPEC) causes the price of oil to rise sharply. High oil prices increase interest in other energy sources.

1999
The Honda Insight becomes the first hybrid electric model in the United States automobile market. The Toyota Prius debuts the following year.

1993
For the first time the United States imports more oil from other countries than it produces.

2000
The world's first commercial wave power station, LIMPET (Land-Installed Marine-Powered Energy Transformer), located on the Scottish island of Islay, begins to generate electricity.

1980
Many wind turbines are installed in California in the early 1980s to help meet growing electricity needs and take advantage of tax incentives.

1970   1980   1990   2000   2007

1978
The Department of Energy's budget for wind power research is $59.6 million; Congress enacts a series of tax benefits to ethanol producers and blenders, which encourage the growth of ethanol production.

1995
Nuclear power contributes about 20 percent of America's electricity.

1998
Iceland unveils a plan to create the world's first hydrogen economy by 2030.

2004
The Energy Information Administration estimates that only about 6 percent of America's energy comes from renewable sources.

1986
The world's worst nuclear power accident happens at the Chernobyl plant in the former USSR (now Ukraine).

2007
President George W. Bush sets the goal of reducing U.S. gasoline consumption by 20 percent in the next 10 years.

Related Organizations

American Council for an Energy-Efficient Economy (ACEEE)

1001 Connecticut Ave. NW, Suite 801

Washington, DC 20036

phone: (202) 429-8873 • fax: (202) 429-2248

e-mail: info@aceee.org • Web site: www.aceee.org

The American Council for an Energy-Efficient Economy is a nonprofit organization dedicated to advancing energy efficiency as a means of promoting both economic prosperity and environmental protection. ACEEE fulfills its mission by conducting in-depth technical and policy assessments, advising policy makers, and working collaboratively with businesses, public interest groups, and other organizations. It also organizes conferences and workshops and publishes numerous reports and books.

American Petroleum Institute (API)

1220 L St. NW

Washington, DC 20005

phone: (202) 682-8000

Web site: www.api.org

The American Petroleum Institute is a trade association representing America's oil and natural gas industry. Its members include producers, refiners, suppliers, pipeline operators, and transporters. API's activities include lobbying, conducting research, and setting technical standards for the petroleum industry.

American Solar Energy Society (ASES)

2400 Central Ave., Suite A

Boulder, CO 80301

phone: (303) 443-3130

Web site: www.ases.org

ASES is a national organization whose mission is to attain a sustainable U.S. energy economy. It is working toward this goal by promoting the development of solar and other renewable energy sources. The organization engages in advocacy, education, research, and encourages collaboration among professionals, policy makers, and the public.

American Wind Energy Association (AWEA)

1101 Fourteenth St. NW, 12th Floor

Washington, DC 20005

phone: (202) 383-2500 • fax: (202) 383-2505

Web site: www.awea.org

AWEA is the national trade association for the wind energy industry and promotes wind energy as a clean source of energy for consumers around the world. The organization provides statistics and information on wind energy and hosts WINDPOWER, a yearly wind energy conference.

Energy & Environmental Research Center (EERC)

University of North Dakota

15 North 23rd St., Stop 9018

Grand Forks, ND 58202

phone: (701) 777-5000 • fax: (701) 777-5181

Web site: www.eerc.und.nodak.edu

The Energy & Environmental Research Center is a nonprofit organization dedicated to developing more efficient energy technologies that also preserve the environment. EERC conducts research, development, and demonstration, and is dedicated to moving promising technologies out of the laboratory and into the commercial marketplace.

Energy Information Administration (EIA)

1000 Independence Ave. SW

Washington, DC 20585

phone: (202) 586-8800

e-mail: infoCtr@eia.doe.gov • Web site: www.eia.doe.gov

The Energy Information Administration is a statistical agency of the U.S. Department of Energy. Its mission is to provide energy data, forecasts, and analyses to promote sound policy making, efficient markets, and public understanding regarding energy and its interaction with the economy and the environment. EIA issues a wide range of weekly, monthly, and annual reports on various energy topics.

International Association for Hydrogen Energy (IAHE)

PO Box 248266

Coral Gables, FL 33124

phone: (305) 284-4666

Web site: www.iahe.org

The IAHE is a group of scientists and engineers professionally involved in the production and use of hydrogen. It believes that hydrogen can be a source of abundant, clean energy and works toward that goal by stimulating the exchange of information about hydrogen energy. The organization publishes a scientific journal and sponsors international exhibitions and conferences.

National Resources Defense Council (NRDC)

40 W. 20th St.

New York, NY 10011

phone: (212) 727-2700

e-mail: nrdcinfo@nrdc.org • Web site: www.nrdc.org

The National Resources Defense Council is a nonprofit activist group that promotes protection of the environment and the use of environmentally safe energy sources. It publishes numerous articles and reports on energy alternatives including wind, solar power, and ethanol.

Nuclear Energy Institute (NEI)

1776 I St. NW, Suite 400

Washington, DC 20006

phone: (202) 739-8000 • fax: (202) 785-4019

e-mail: webmasterp@nei.org • Web site: www.nei.org

The Nuclear Energy Institute is a policy organization of the nuclear energy industry. It has over 280 corporate members in 15 countries. The organization promotes policies that will increase the beneficial uses of nuclear energy technologies in the United States and around the world.

Renewable Fuels Association (RFA)

1 Massachusetts Ave. NW, Suite 820

Washington, DC 20001

phone: (202) 289-3835 • fax: (202) 289-7519

e-mail: info@ethanolrfa.org • Web site: www.ethanolrfa.org

RFA is the national trade association for the U.S. ethanol industry. Its membership includes businesses, individuals, and organizations dedicated to expanding the use of ethanol fuel in the United States. It promotes policies, regulations, and research that will lead to increased production and use of ethanol fuel.

Union of Concerned Scientists (UCS)

2 Brattle Sq.

Cambridge, MA 02238

phone: (617) 547-5552 • fax: (617) 864-9405

e-mail: ucs@ucsusa.org • Web site: www.ucsusa.org

The Union of Concerned Scientists is a nonprofit alliance of scientists who argue that energy alternatives to oil must be developed to reduce pollution and slow global warming. It combines scientific research and citizen action in an effort to change government policy, corporate practices, and consumer choices. UCS publishes numerous articles and reports on alternative energy sources.

For Further Research

Books

Godfrey Boyle, ed., *Renewable Energy.* New York: Oxford University Press, 2004.

Kristen A. Day, ed., *China's Environment and the Challenge of Sustainable Development.* Armonk, NY: M.E. Sharpe, 2005.

Kenneth S. Deffeyes, *Beyond Oil: The View from Hubbert's Peak.* New York: Hill and Wang, 2005.

Howard Geller, *Energy Revolution: Policies for a Sustainable Future.* Washington, DC: Island, 2003.

Barry J. Hanson, *Energy Power Shift: Benefiting from Today's New Technologies.* Maple, WI: Lakota Scientific, 2004.

Richard Heinberg, *The Party's Over: Oil, War, and the Fate of Industrial Societies.* Gabriola Island, BC: New Society, 2003.

James Howar, *The Long Emergency: Surviving the Converging Catastrophes of the Twenty-First Century.* New York: Atlantic Monthly, 2005.

Peter W. Huber, *The Bottomless Well: The Twilight of Fuel, the Virtue of Waste, and Why We Will Never Run Out of Energy.* New York: Basic Books, 2005.

Jeremy Leggett, *The Empty Tank: Oil, Gas, Hot Air, and the Coming Global Financial Catastrophe.* New York: Random House, 2005.

Karl Mallon, ed., *Renewable Energy Policy and Politics: A Handbook for Decision-Making.* Sterling, VA: Earthscan, 2006.

Amos Salvador, *Energy: A Historical Perspective and 21st Century Forecast.* Tulsa, OK: American Association of Petroleum Geologists, 2005.

Herman Scheer, *The Solar Economy: Renewable Energy for a Sustainable Global Future.* London: Earthscan, 2004.

William Sweet, *Kicking the Carbon Habit: Global Warming and the Case for Renewable and Nuclear Energy.* New York: Columbia University Press, 2006.

Periodicals

Mark Baard, "Hydrogen's Dirty Details," *Village Voice*," January 7–13, 2004.

Joel Bainerman, "The Myths and Hype of Hydrogen," *Middle East,* May 2006.

Dennis Behreandt, "The Promise of Synthetic Fuel," *New American,* November 27, 2006.

Robert L. Bradley Jr., "Are We Running Out of Oil? 'Functional Theory' Says No," *PERC Reports,* September 2004. www.perc.org.

Business Week, "Green Homes: The Price Still Isn't Right," February 12, 2007.

Colin J. Campbell, "Peak Oil: A Turning Point for Mankind," *Social Contract,* Spring 2005.

Mike Clowes, "Cornfed, Not Corn Fueled Is the Way to Go," *Investment News,* February 5, 2007.

Commercial Motor, "A World Without Oil," January 25, 2007.

Economist, "Blowing Hot and Cold; Geothermal Energy," September 16, 2006.

———, "Canola and Soya to the Rescue; Alternative Energy," May 6, 2006.

Joseph Florence, "Wind Power Blowing Harder," *USA Today* (magazine), November 2006.

Danylo Hawaleshka, "Power Hungry," *Maclean's,* February 25, 2005.

Bob Herbert, "Oil and Blood," *New York Times,* July 28, 2005.

Kimberly Lisagor, "Sunshine's Bottom Line," *Mother Jones,* January/February 2007.

Amory B. Lovins, "How to Live Without Oil," *Newsweek,* August 8, 2005.

Michelle Nijhuis, "Selling the Wind," *Audubon,* September/October 2006.

Mike Parfit, "After Oil: Powering the Future," *National Geographic,* August 2005.

Eric Scigliano, "Wave Energy," *Discover,* December 2005.

Jerry Taylor, "For Now, Gasoline Is Our Only Cheap Fuel," *Arizona Republic,* May 7, 2006.

Jerry Taylor and Peter Van Doren, "Stuck on Empty," *National Review,* February 3, 2006.

Matthew L. Wald, "Wind Power Is Becoming a Better Bargain," *New York Times,* February 13, 2005.

George F. Will, "Inconvenient Kyoto Truths," *Time,* February 12, 2007.

Drew Winter, "The Case for Hydrogen," *Ward's Auto World,* June 1, 2006.

Internet Sources

Energy Information Administration, "Renewable Energy Sources: A Consumer's Guide," December 1, 2005. www.eia.doe.gov/neic/brochure/renew05/renewable.html.

National Resources Defense Council, "Ethanol: Energy Well Spent," 2006. www.nrdc.org/air/transportation/ethanol/ethanol.pdf.

———, "Wind, Solar, and Biomass Energy Today," January 12, 2006. www.nrdc.org/air/energy/renewables/overview.asp.

David Sandalow, "Ethanol: Lessons from Brazil," Brookings Institution, May 2006. www.brookings.edu/views/articles/fellows/sandalow_20060522.pdf.

White House National Economic Council, "Advanced Energy Alternative," February 2006. www.whitehouse.gov/stateoftheunion/2006/energy/energy_booklet.pdf.

Source Notes

Overview

1. United Nations Development Program, United Nations Department of Economic and Social Affairs, and the World Energy Council, "World Energy Assessment: Overview—2004 Update," 2004. www.undp.org.

2. J. Andrew Hoerner and James Barrett, "Smarter, Cleaner, Stronger: Secure Jobs, a Clean Environment, and Less Foreign Oil," *Redefining Progress,* October 2004. www.redefiningprogress.org.

3. Howard Geller, *Energy Revolution: Policies for a Sustainable Future.* Washington, DC: Island, 2003, p. 1

4. Jeremy Leggett, *The Empty Tank.* New York: Random House, 2005, p 3.

5. Sierra Club, "Global Warming: A Time for Action." www.sierraclub.org.

6. World Coal Institute, "Coal: Secure Energy," October 2005. www.world coal.org.

7. M.A. Adelman, "The Real Oil Problem," *Regulation,* Spring 2004, p. 17.

8. Energy Information Administration, "Energy in the United States: 1635–2000." www.eia.doe.gov.

9. Energy Information Administration, "Energy in the United States."

10. iTulip, "Energy and Money Part I: Too Little Oil or Too Much Money?" *Weekly Commentary,* May 4, 2006. www.itulip.com.

11. Jerry Taylor, "For Now, Gasoline Is Our Only Cheap Fuel," *Arizona Republic,* May 7, 2006. www.azcentral.com.

12. National Resources Defense Council, "Energy Facts," January 2007. www. nrdc.org.

13. Sarah Barr, "Update: Alternative Energy in the United States," *OneWorldUS,* November 2006. http://us.oneworld.net.

14. National Resources Defense Council, "A Responsible Energy Plan for America," April 2005. www.nrdc.org.

15. Karl Mallon, "Introduction," in *Renewable Energy Policy and Politics: A Handbook for Decision-Making,* ed. Karl Mallon. Sterling, VA: Earthscan, 2006, p. 2.

16. Dennis Behreandt, "Energy's Future," *New American,* April 4, 2005. www. thenewamerican.com.

17. Barry J. Hanson, *Energy Power Shift: Benefiting from Today's New Technologies.* Maple, WI: Lakota Scientific, 2004, p. 8.

18. International Energy Agency, *World Energy Outlook 2006.* Paris, France: International Energy Agency, 2006, p 37.

19. White House National Economic Council, "Advanced Energy Alternative," February 2006. www.whitehouse. gov.

20. Thomas C. Dorr, testimony before the House Agriculture Committee, June 29, 2006.

Are Alternative Energy Sources Necessary?

21. International Energy Agency, *World Energy Outlook 2006,* p. 37.

22. Leggett, *The Empty Tank,* p. 5.

23. Colin J. Campbell, "Peak Oil: A Turning Point for Mankind," *Social Contract,* Spring 2005, p. 182.

24. Robert L. Bradley Jr., "Are We Running Out of Oil? 'Functional Theory' Says No," *PERC Reports,* September 2004. www.perc.org.

25. White House National Economic Council, "Advanced Energy Alternative."

26. Geller, *Energy Revolution,* p. 11.

27. Marty Bender, "What Will Come After Fossil Fuels?" *Land Institute,* August 21, 2003. www.landinstitute.org.

28. Leggett, *The Empty Tank,* pp. xiv–xv.

29. Leggett, *The Empty Tank,* p 5.

30. Chuck Alston, "Natural Gas: Bridge to a Clean Energy Future," *Progressive Policy Institute,* June 6, 2003. www.ppionline.org.

31. National Energy Technology Laboratory, "Technologies: Oil and Natural Gas Supply." www.netl.doe.gov.

32. Raymond J. Kopp, "Natural Gas: Supply Problems Are Key," *Resources for the Future,* Winter 2005. www.rff.org.

33. David Talbot, "The Dirty Secret," *Technology Review,* July 19, 2006. www.technologyreview.com.

34. World Coal Institute, "Clean Coal: Building a Future Through Technology," 2004. www.worldcoal.org.

35. Leggett, *The Empty Tank,* p. 168.

36. Quoted in Brian Braiker, "Crude Awakening," *Newsweek,* February 17, 2004. www.msnbc.msn.com.

37. National Resources Defense Council, "A Responsible Energy Plan for America."

38. Dallas Burtraw and Karen L. Palmer, "Cleaning Up Power Plant Emissions," in *New Approaches on Energy and the Environment: Policy Advice for the President,* eds. Richard D. Morgenstern and Paul R. Portney. Washington, DC: Resources for the Future, 2004, p. 47.

39. National Center for Policy Analysis, "HS Debate: Energy." www.ncpa.org.

40. Leggett, *The Empty Tank,* pp. 79-80.

41. George F. Will, "Inconvenient Kyoto Truths," *Newsweek,* February 12, 2007, p. 72.

42. Mallon, "Introduction," in *Renewable Energy Policy and Politics,* p. 1.

What Alternative Energy Sources Should Be Pursued?

43. Hanson, *Energy Power Shift,* p. 15.

44. Patrick Moore, testimony before the Senate Committee on Energy and Natural Resources, April 28, 2005.

45. Moore, testimony.

46. Bill McKibben, "One Roof at a Time," *Mother Jones,* November/December 2004. www.motherjones.com.

47. James, "Comments to 'Solar Power Too Cheap to Meter?'" *Reason,* February 21, 2007. www.reason.com.

48. Quoted in Brian Braiker, "Crude Awakening."

49. Kimberly Lisagor, "Sunshine's Bottom Line," *Mother Jones,* January/February 2007.

50. Jay Lehr, "Solar Power: Too Good to Be True," *Heartland Institute,* June 1, 2005. www.heartland.org.

51. White House National Economic Council, "Advanced Energy Alternative."

52. H. Sterling Burnett, "Wind Power: Red, Not Green," *Brief Analysis No. 467,* National Center for Policy Analysis, February 23, 2004. www.ncpa.org.

53. Michelle Nijhuis, "Selling the Wind: Wind Power Is Pollution-Free, Combats Global Warming, and Is a Boon to Small Farmers. The Biggest Drawback—Its Lethal Impact on Birds and Bats—Is Driving Creative Ways to Ensure That This Fast-Growing Energy Source Can Coexist with Wildlife," *Audubon,* September/October 2006.

54. National Resources Defense Council, "Wind, Solar, and Biomass Energy Today," January 12, 2006. www.nrdc.org.

55. Quoted in Nijhuis, "Selling the Wind."

56. Quoted in Nijhuis, "Selling the Wind."

57. Energy Information Administration, "Hydropower—Energy from Moving Water," *Energy Kid's Page,* February 2006. www.eid.doe.gov.

58. Environmental Protection Agency,

"Electricity from Hydropower," July 19, 2006. www.epa.gov.

59. U.S. Department of Energy, "Geothermal FAQs," January 13, 2006. www1.eere.energy.gov.
60. Karl Gawell and Diana Bates, "Geothermal Literature Assessment: Environmental Issues," May 2004. www.geothermal-biz.com.
61. Eric Scigliano, "Wave Energy," *Discover,* December 2005. www.discover.com.
62. Massachusetts Technology Collaborative, "Ocean Energy (Wave, Tidal, Ocean Thermal)." www.mtpc.org.
63. Michael Parfit, "Future Power: Where Will the World Get Its Next Energy Fix?" *National Geographic,* August 2005. www.nationalgeographic.com.

Can Alternative Energy Be Used for Transportation?

64. White House National Economic Council, "Advanced Energy Alternative."
65. Drew Winter, "The Case for Hydrogen," *Ward's Auto World,* June 1, 2006.
66. White House National Economic Council, "Advanced Energy Alternative."
67. Joel Bainerman, "The Myths and Hype of Hydrogen," *Middle East,* May 2006.
68. Bainerman, "The Myths and Hype of Hydrogen."
69. Chris Demorro, "Prius Outdoes Hummer in Environmental Damage," *Recorder,* March 7, 2007. http://clubs.ccsu.edu.
70. Daniel, "Kluger Hybrid Still on Hold," *CarsGuide.com.au,* February 23, 2007. http://carsguide.news.com.au.
71. Hybrid Electric Cars, "12 Common Hybrid Car Myths." http://hybrid-electric-car.net.
72. Peter Valdes-Dapena, "Hybrids: Seven Worries, Seven Answers," *CNNMoney.com,* March 1, 2007. http://money.cnn.com.
73. Christopher Todd, "My Civic Hybrid Experience!" March 18, 2007. www.gaianar.com.
74. Alexander Karsner, testimony before the Committee on Environment and Public Works, U.S. Senate, September 6, 2006.
75. National Resources Defense Council, "Wind, Solar, and Biomass Energy Today."
76. International Energy Agency, *World Energy Outlook 2006,* p. 413.
77. Glen Barry, "Bursting Biofuels' Bubble," *Earth Meanders,* April 25, 2006. http://earthmeanders.blogspot.com.
78. George Monbiot, "Fuel for Nought," *Guardian,* November 23, 2004. www.guardian.co.uk.
79. Josh Tickell, interview by Libby Tucker, "Biodiesel America," *Daily Journal of Commerce, Portland,* February 9, 2007.
80. Mike Clowes, "Cornfed, Not Corn Fueled Is the Way to Go," *Investment News,* February 5, 2007.
81. Union of Concerned Scientists, "Ethanol: Frequently Asked Questions," July 7, 2006. www.ucsusa.org.
82. David Sandalow, "Ethanol: Lessons from Brazil," May 2006. www.brookings.edu.
83. André Kenji de Sousa, "There Is No Ethanol Revolution in Brazil," *Lew Rockwell.com,* August 9, 2006. www.lewrockwell.com.
84. Tickell, "Biodiesel America."

How Will Increased Use of Energy Alternatives Impact Society?

85. Ben Hewitt, "The Hidden Cost of Biofuels," *Popular Mechanics,* December 6, 2006. www.popularmechanics.com.
86. Hanson, *Energy Power Shift,* p. 15.
87. Richard Heinberg, *The Party's Over: Oil, War, and the Fate of Industrial Societies.*

Gabriola Island, BC: New Society, 2003, p. 201.

88. United Nations Development Program, United Nations Department of Economic and Social Affairs, and the World Energy Council, "World Energy Assessment.

89. John Flicker, "Audubon View (Wind Power)," *Audubon,* November/December 2006.

90. National Center for Policy Analysis, "HS Debate: Energy."

91. Bill Peacock, "Renewable Energy's Real Cost," *Statesman.com,* January 31, 2006. www.statesman.com.

92. Geller, *Energy Revolution,* p. 12.

93. Barack Obama, remarks to the Governor's Ethanol Coalition, Washington, D.C., February 28, 2006.

94. United States Energy Association, "National Energy Security Post 9/11," June 2002. www.usea.org.

95. Hoerner and Barrett, "Smarter, Cleaner, Stronger."

96. Michael Kane, "Renewable Energy: Too Little, Too Late," *Guerrilla News Network,* October 12, 2005. www.gnn.tv.

97. Kane, "Renewable Energy."

98. American Solar Energy Society, "Frequently Asked Questions." www.ases.org.

99. Jim McKay, "Renewable Energy Still May Be Too Expensive," *Pittsburgh Post-Gazette,* October 23, 2005. www.post-gazette.com.

100. Union of Concerned Scientists, "The Hidden Costs of Fossil Fuels," August 10, 2005. www.ucsusa.org.

101. Union of Concerned Scientists, "The Hidden Costs of Fossil Fuels."

102. William von Brethorst, response to "Are Sun and Wind Energy the Answer?" *Principal Voices,* 2006. www.principalvoices.com.

103. Schatz Energy Research Center, "Renewable Energy FAQs." www.humboldt.edu.

104. Jim Holm, "Alternative Energy Sources Are Not Alternatives to Oil," *Nuclear oil.com,* April 3, 2007. www.nuclear oil.com.

105. Cheryl Coon, "Wind Power: Benevolent Solution or Hidden Danger?" Audubon Society of Portland. www.audubonportland.org.

List of Illustrations

Index

About the Author

Andrea C. Nakaya, a native of New Zealand, holds a BA in English and an MA in Communication from San Diego State University. She currently lives in Encinitas, California, with her husband Jamie and their daughter Natalie. In her free time she enjoys traveling, reading, gardening, and snowboarding.